Recommended Citation: Glauber, J., and D. Laborde, eds. 2023. *The Russia-Ukraine Conflict and Global Food Security*. Washington, DC: International Food Policy Research Institute.

International Food Policy Research Institute
1201 Eye Street, NW
Washington, DC 20005-3915 USA
www.ifpri.org

ISBN: 978-0-89629-439-4
DOI: https://doi.org/10.2499/9780896294394

Photo credits
COVER: Leah Millis / Reuters.
CHAPTER IMAGES: Pg. 13, Homes for heroes, Shutterstock.com.

Design and layout: Lee Dixon
Project manager: Pamela Stedman-Edwards

TABLE OF CONTENTS

Acknowledgments

The research undertaken for much of this e-book was supported by grants from the United Kingdom's Foreign, Commonwealth and Development Office (FCDO) and the United States Agency for International Development (USAID). We thank all the IFPRI authors and external collaborators for their contributions and IFPRI's Communications & Public Affairs Unit for its support to this body of work. The authors and collaborators are also thankful for research and data inputs provided by IFPRI's Food Security Portal, a knowledge platform supported with funding of the European Union (EU).

Foreword

The Russian invasion of Ukraine in February 2022 created an immediate shock to global markets. Prices for cereals and vegetable oils rose rapidly in expectation of disruptions to Ukraine's harvests and to shipping through the Black Sea. The war exacerbated existing economic fragility and food insecurity caused by climate change, conflict, and COVID-19, and the reaction of the international community to spikes in fuel and fertilizer prices led to additional impacts on agricultural markets. For many countries, a serious food crisis seemed likely.

In the face of this sudden shock, policymakers needed information and guidance. IFPRI, with support provided by USAID and FCDO, stepped up immediately with analysis of global markets and country-level analysis, providing both a picture of how this crisis would play out and how (and how not) to address the threat of growing food insecurity and rising prices. This work has drawn on IFPRI's many years of research as well as its strong presence in many developing regions.

Most importantly, IFPRI strove to get this information to stakeholders who were trying to make sense of the myriad implications of the Russia-Ukraine conflict – policymakers, the press, funding organizations, NGOs, and others. This book compiles 35 blogposts written over the past year – beginning the very day of the invasion – on Food and Fertilizer Price Trends. These posts, featured on IFPRI's webpage on the food price crisis, have garnered almost 180,000 unique page views, including at least 7 that have been viewed more than 7,000 times and one that has been viewed over 39,000 times. IFPRI also hosted 7 policy seminars between July and October, bringing together experts and researchers to discuss the impact of the crisis and explore responses, which attracted close to 8,000 live views and continue to be viewed or listened to as podcasts.

IFPRI also released a set of 20 publications based on modeling of the crises impacts and potential context-appropriate policy responses in a range of developing countries, which have already been downloaded more than 2,500 times.

In addition, IFPRI has developed a number of tools to help practitioners navigate the food crises, including a Food and Fertilizer Export Restrictions Tracker, an Excessive Food Price Variability Early Warning System, and a Fertilizer Dashboard. The Food Security Portal, which hosts these tools and provides access to other data on food systems, was viewed more than 33,000 times in 2022.

The success in communicating this work is most evident in the media attention it has received. IFPRI's work on the food price crisis has been cited in top newspapers, magazines, and web-based news sources, including more than five mentions in The Washington Post, Financial Times, CNBC, Bloomberg, Wall Street Journal, New York Times, Yahoo, and the Economist, as well as mentions in The Conversation, Associated Press, Reuters, AllAfrica, The Business Standard (Bangladesh), Dhaka Tribune, The Hindu (India), Voice of America, Los Angeles Times, and Devex, as well as thousands of reposts.

This book brings together IFPRI's blogposts about the Ukraine crisis, written by IFPRI researchers and their colleagues. They cover the broad range of impacts and policy responses to the crisis. It is our hope that these will continue to inform the management of the present crisis and of future crises that disrupt our food systems.

Charlotte Watts *Chief Scientific Adviser and Director for Research and Evidence, United Kingdom Foreign, Commonwealth & Development Office*

Robert Bertram *Chief Scientist, Bureau for Resilience and Food Security, United States Agency for International Development*

Johan Swinnen *Director General, IFPRI, and Managing Director, Systems Transformation Science Group, CGIAR*

INTRODUCTION

When Russia invaded Ukraine in February 2022, the conflict quickly sparked fears of a global food crisis. Food prices were already high in the wake of the COVID-19 pandemic, and many countries were facing serious food insecurity. The Black Sea region is critical to meeting the world's food needs, as is Russia's role in supplying natural gas and, along with Belarus, fertilizers. Any disruption to this trade would drive up food prices and aggravate poverty, hunger, and malnutrition in vulnerable countries. With other countries undertaking export restrictions, these trade disruptions became even larger. The conflict has primarily affected food systems through effects on trade in cereals and vegetable oils, interruptions to fertilizer exports, and rising energy prices, although a drop in agricultural production in Ukraine is also a significant factor. Together, these impacts could drive millions more people into hunger and poverty. Early in the crisis, the FAO estimated that a prolonged disruption of exports from Ukraine and Russia could increase the number of undernourished people by 8 to 13 million in 2022.

IFPRI responded rapidly to the need for information and policy advice to address this new crisis. We signaled the worrying outlook for some agricultural commodities prior to the onset of conflict, and our first analysis on potential impacts was released on February 24, the day of the invasion. Mindful of the urgent need for analysis, IFPRI established a new blog series on High Food and Fertilizer Prices and War in Ukraine. This series, which was supported by a set of online trackers and dashboards providing access to detailed information, allowed for rapid dissemination of critical information and was widely shared through our website and communications with policymakers and media outlets. IFPRI's strong presence at the country level allowed researchers to provide important insights into the conflict's impact on vulnerable countries and regions. The blog posts include timely analysis of trade flows, tracking of food prices and policy responses, and results of impact modeling for vulnerable countries, which were more fully examined in a series of briefs.

This book is a compilation of those blog posts. Together, they provide an overview of how the crisis has progressed, how the international community and individual countries responded with efforts to ensure food security, and what we are learning about the best ways to ensure food security in the aftermath of a major shock to global food systems. These blog posts each reflect a moment in time – the Russia-Ukraine war and its global repercussions have continued to evolve, especially since the earliest posts were written. Thus, they provide a vivid set of "snapshots" that can help us understand and learn from this crisis. As of this writing, international prices have declined from their post-invasion highs, but as the crisis continues to unfold, further volatility seems almost certain. Moreover, compounding factors have exacerbated inflation around the world. Given that the total supply of both food and fertilizers may be enough for all of humanity, the crisis remains one of affordability rather than availability.

Role of Russia and Ukraine in global markets

Russia and Ukraine both play a key role in world food markets, so any disruption in the region has global implications. In recent decades, the Black Sea region has become a major supplier of grains, oilseeds, and vegetable oil. Prior to the conflict, Russia and Ukraine accounted for 12% of total calories traded in the world and 30% of global wheat and barley exports. They were among the top five global exporters of wheat, barley, sunflower seeds, and maize. In addition, Ukraine supplied about half of the global market for sunflower oil. Many low- and middle-income countries, especially in the Middle East and North Africa, rely heavily on imports of wheat, barley, and maize from Russia and Ukraine. Europe and China are also major importers of Ukraine's feed products. These commodities play a vital role in diets, both directly (wheat, vegetable oils) and indirectly through the livestock industry (maize, barley, but also wheat and sunflower meals). Ukraine's exports were severely disrupted as a direct result of the conflict. Russia's exports of agricultural commodities and fertilizers were also affected in the aftermath of the invasion, due to concerns on the part of some companies about running afoul of financial sanctions imposed on Russia, but subsequently rebounded. Export restrictions imposed by other countries after the invasion also contributed to the dramatic increases in global prices, which posed a threat to food security and nutrition for many countries.

Beyond food products, key inputs to the food system are also affected. Russia is a major source of natural gas, exporting about 20% of globally traded natural gas and, before the war, supplying some 40% of the European Union's imports. Along with its ally Belarus, Russia is an important exporter of potash, phosphate, and nitrogenous fertilizers (ammonia and urea), for which natural gas is a critical input. Ukraine and many other countries, including many in Europe, sub-Saharan Africa, and Latin America, have been heavily dependent on these fertilizer supplies. Rising energy prices increase not only the cost of fertilizer production, but also the cost of food processing and transport. Rising fertilizer prices contribute to rising food prices and, especially in countries that are not seeing concomitant increases in output prices, they also cut into farmers' profits and income, reduce fertilizer use, and consequently lower productivity. These problems are magnified for smallholders already facing various market failures, especially difficulties in accessing finance, and they risk being completely excluded from accessing inputs.

Price impacts of the conflict

In late 2021, prices for food, including staple grains and vegetable oils, reached their highest levels in a decade. Rising prices were driven by supply chain issues related to the COVID-19 pandemic, as well as poor harvests in North and South America and Malaysia and increased use for biodiesel production (Figure 1). Low- and

Figure 1: Real prices for food, fertilizer, and energy (index)

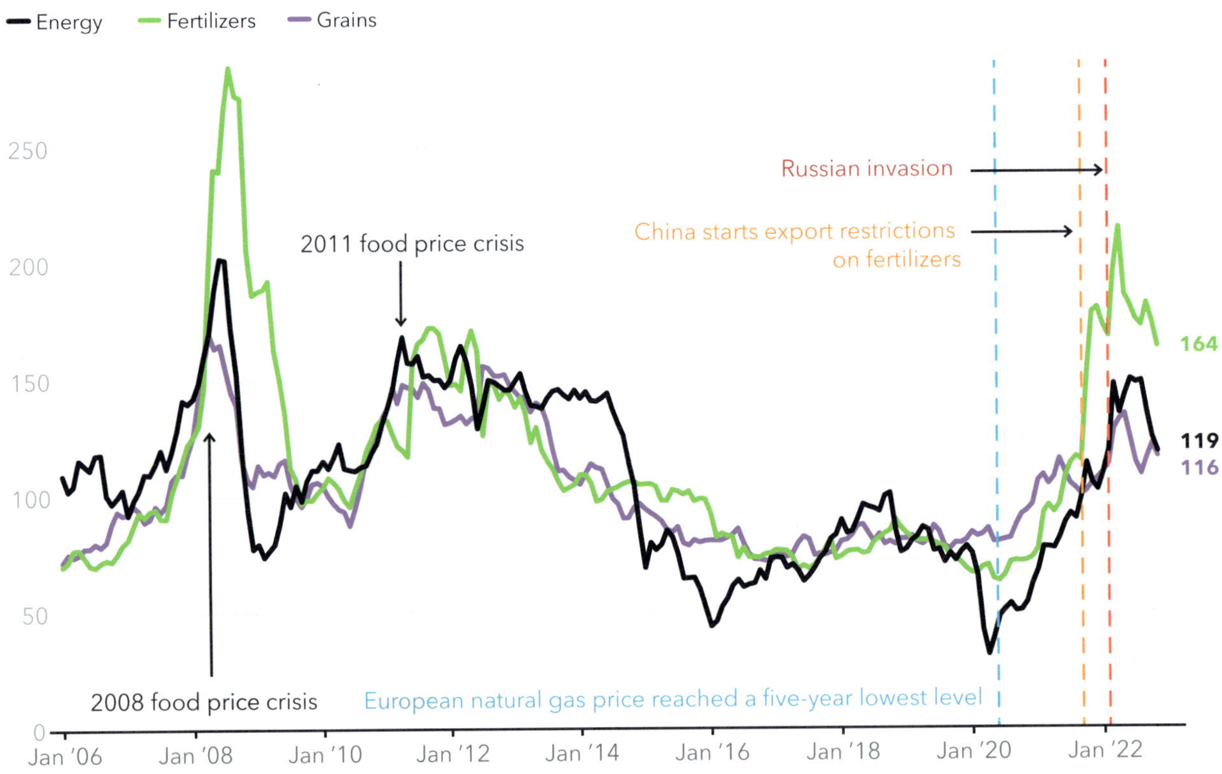

Source: World Bank, U.S.Bureau of Labor Statistics

middle-income countries, including not only net food importers but also exporting countries such as Brazil, were already experiencing increased malnutrition as a result of these multiple crises.

Prices and price volatility for agricultural commodities, fuel, and fertilizers increased immediately in the wake of Russia's February invasion of Ukraine. Between 2019 and March 2022, cereal prices rose by 48%, fuel prices by 86%, and fertilizer prices by 35%. By July 2022, prices of agricultural commodities had largely returned to February 2022 levels, defying some predictions. Yet, these prices are high by historical measures, and food prices remain well above pre-COVID-19 levels, while more generalized inflation is hitting consumers and governments hard. The disruption of Ukraine's exports added to other significant pressures on grain and oilseed markets. Starting in 2021, the scarcity of supplies impacted all major crops except rice. La Niña shifted weather patterns around the globe, leading to drought in important wheat-exporting areas, notably North America, as well as in Brazil and Paraguay, where it affected the soybean crop and subsequently reduced soybean oil production. In this context, the inventory of these products was low and any new shocks triggered price spikes.

International responses

The international community has responded to the conflict with economic and financial sanctions on Russia, adding to those already imposed after its annexation of Crimea in 2014. The sanctions, meant to limit Russia's capacity to fund the war, clearly exempt food and fertilizers. However, sanctions may have had an indirect impact in these sectors in the early months of the conflict, as some companies may have been concerned by the financial sanctions. In addition, Russia imposed its own export restrictions. While Russia's exports decreased in the first months after the invasion, its total 2022 exports of grains and fertilizers (with the exception of ammonia) were in line with previous years and export earnings increased as a result of the higher prices.

Given Ukraine's important role as a major breadbasket, negotiations sought to protect these exports. Yet Russia's blockade of Ukraine's Black Sea ports, plus Ukrainian mines placed to forestall Russian attack, effectively prevented cargo ships from leaving these ports. An agreement between Russia and Ukraine, brokered by Türkiye and the United Nations in July 2022, allowed for exports from the ports, under Ukrainian escort, to relieve the bottleneck of Ukrainian wheat. However, Russia's brief withdrawal from the agreement in October provided a reminder of how tenuous the situation remains.

The international community has also responded with financial and knowledge support for vulnerable countries, including a G7-led Global Alliance for Food Security and a UN Global Crisis Response Group. Financial support for agriculture and food security in the most vulnerable countries is expected to increase by about US$12 billion. Members have committed to increasing support for the G20's Agricultural Market Information System (AMIS), which provides greater transparency on food, agriculture, and, increasingly, fertilizer markets, helping to limit disruptions. Indeed, evidence-based and accurate information related to crises is key to reassuring markets and avoiding ill-advised policies.

In a misguided attempt to guard their national food security, several countries imposed restrictions on exports of key commodities. Export bans enacted to guard food security during the COVID-19 pandemic were generally short-lived, but the Russia-Ukraine war revived interest in them. By mid-April 2022, 17% of total traded calories were affected by trade bans. By December 2022, this level had fallen to 7.3%, but unfortunately, such efforts to insulate domestic markets can have negative consequences for the most vulnerable and contribute to the global food price crisis. They also hurt local farmers, as demonstrated by the protests and concerns raised by smallholders in India following the export bans on wheat and in Indonesia following the ban on palm oil exports.

Country impacts

The impacts of the Russia-Ukraine war and related food price volatility vary by country and region, as do the appropriate responses to ensure food security for their populations. The Middle East and North Africa region (MENA) has been most directly affected by the conflict. MENA is the largest importer of Russian and Ukrainian wheat, and also a major importer of other grains and vegetable oil from Ukraine. Egypt, for example, is a major consumer of wheat and heavily subsidizes bread for its consumers. The country imports 62% of its supply, primarily from Russia and Ukraine, making it the world's largest wheat importer. For countries like this, the short-term solution lies in finding new suppliers and readjusting social safety nets to better target the neediest and to support consumption of other foods. In the longer term, such countries will need to pursue several options, including increasing sustainable domestic production through investment in R&D and infrastructure, reducing wheat consumption and increasing dietary diversity (with the potential added benefit of healthier diets), and better targeting their social safety nets.

Several countries in the MENA region face a "polycrisis" – not only slow economic recovery and high food prices in the wake of COVID-19, but also protracted war or civil strife, economic collapse, and long-term climate challenges. This is the case for Yemen, Sudan, Lebanon, and other countries where the food and fuel price crisis is an added stressor in an already dire situation. In Yemen, where civil war has been ongoing since 2015, the economy is devastated, and poverty and malnutrition are pervasive. Half of the population receives food assistance, and imported wheat and vegetable oil are a major source of calories. For these countries, an increase in humanitarian aid is the only viable short-term solution. Protecting their commercial imports from export restrictions also remains essential. For the longer term, conflict resolution and macroeconomic recovery are critical to reestablishing food security.

The indirect impacts of the crisis will also be quite significant in many vulnerable countries, especially in Africa south of the Sahara, where rates of poverty and malnutrition are generally high, and governments have limited means. While the global food price increase may have little negative impact on local food sources, such as cassava, the overall rise in prices for imported foods and many consumer goods – resulting from fuel price increases – will affect poor consumers most severely. For example, in Malawi, maize accounts for 36% of food purchases by the poor, even though many of them produce some of the maize they eat. Any increase in the price of maize will curtail their ability to feed themselves. At the same time, the rise in fertilizer prices will cut into farmers' profits and lead many to limit fertilizer use, thus reducing production and food availability. The short-term response for these countries will require strengthening safety net programs and extension services to help farmers optimize their fertilizer use. Long-term solutions require investment in agricultural production, including R&D and infrastructure to boost sustainable productivity and adapt to climate change. Indeed, mitigating both internal and external shocks calls for a balanced approach. But both of these solutions put governments in a predicament – they require the use of scarce funds, which will likely lead to more pressure on foreign reserves and further increase the price of imports.

What IFPRI has done

In response to the crisis, IFPRI researchers developed and enhanced several tools for both analysts and policymakers, and produced and publicized modeling results and analysis at the global and country level for countries most vulnerable to food system disruptions. Through our Food Security Portal, IFPRI provides access to several useful tools that compile up-to-date information on food systems. The Excessive Food Price Variability Early Warning System identifies periods of excessive price variability for major agricultural crops and is updated on a daily basis. Tracking food price volatility serves as a "temperature check" on global food markets, as well as an important early warning for potential crises. The Fertilizer Dashboard provides several years of data on fertilizer prices and trade, and the Food and Fertilizer Export Restrictions Tracker provides publicly available data on restrictions affecting trade in major food and fertilizer markets.

To predict likely impacts of the crisis, researchers used IFPRI's economywide model (RIAPA) to estimate the effects of food, fuel, and fertilizer price shocks for a wide range of countries in Asia, Africa, and the Middle East. Using country-specific parameters, these simulations examine diverse effects on GDP, employment, production, household consumption, inequality, poverty, and diet quality, which serve as the basis for country-appropriate responses recommended in a series of briefs.

IFPRI is also hosting a policy seminar series on Food and Fertilizer Price Trends, which brings experts together to discuss the potential impacts of rising prices and policy responses on global food security and nutrition.

Lessons from the crisis

Crises such as this have roots in multiple shocks or long-term pressures – in this case, the slow recovery from the global pandemic, conflict, and climate-related disasters – and are becoming increasingly common, especially as climate change advances. These crises not only overlap but also amplify disruptive impacts on food production and markets. Such complex situations will likely drive rising numbers of food-insecure and malnourished people, disrupt farmers' livelihoods and leave long-lasting impacts on well-being. The links between drought, war, and food insecurity, for example, are evident in many places.

Our experience with the current food price crisis offers several key policy lessons, which are elaborated in the blog posts that follow. Foremost among them is the critical role of trade in ensuring food security. Keeping markets open for food and fertilizers – and expanding the number of producers and markets – can reduce price volatility and help ensure delivery of food where it is needed. Grain and vegetable oil supplies can also be increased in the short term by suspending biofuel mandates and avoiding taking land out of food and feed production. In the medium term, countries can invest in increasing sustainable food production. As more countries develop resilient and competitive agricultural systems, importing countries will have access to more trade partners and be able to diversify their sources of imports. Fertilizer production is not easily increased in the short run, but efforts should be pursued to ensure increased efficiency and efficacy of fertilizer application, as well as innovative plant nutrition solutions. Finally, humanitarian assistance for those most in need and well-targeted social protection, through food or cash transfers, can prevent hunger and malnutrition and deter the devastating long-term impacts of a global food crisis, but these should not detract from efforts to meet long-term development goals and build resilience to future shocks.

RUSSIAN-UKRAINE WAR: 2022 TIMELINE OF EVENTS

JAN · FEB · MAR · APR · MAY · JUN

FEBRUARY 24: Russian forces attack Ukraine.

FEBRUARY/MARCH: The United States, European Union, and allies issue economic sanctions on Russia.

MARCH 2: The UN General Assembly condemns Russia's invasion of Ukraine.

MARCH: The war displaces nearly one-quarter of Ukraine's population.

MARCH: Staple food prices grow more volatile, and the FAO warns that the conflict could threaten global food security.

APRIL: Prices for fertilizer, grains, and oilseed, and energy rise sharply.

APRIL: Russia withdraws all troops from Kyiv before launching a new offensive in eastern Ukraine.

MAY: Russian forces take control of Mariupol, a highly strategic southeastern port city.

JUL — AUG — SEP — OCT — NOV — DEC

JUNE/JULY: The EU adopts a plan to limit its dependence on Russian energy.

EU and G7 countries issue new sanctions and expand the SWIFT ban.

JULY: Russia and Ukraine agree to the Black Sea Grain Initiative to allow food and fertilizer exports.

AUGUST: Shelling near Ukraine's Zaporizhzhia nuclear power plant renews fears of a nuclear accident.

SEPTEMBER: Ukrainian forces make major advances, and Russia moves to annex occupied territory.

SEPTEMBER: FAO's Food Price Index declines substantially, but food price inflation remains high globally.

OCTOBER: Russia renews its offensive with missile strikes in 14 regions of Ukraine.

NOVEMBER: The UN reports more than 6,000 civilian deaths and 10,000 injuries from the war. Nearly 8 million people have fled as refugees and more than 6.5 million are internally displaced.

NOVEMBER: The Black Sea Grain Initiative is extended for 120 days.

SECTION 1

A CONFLICT WITH GLOBAL CONSEQUENCES

1 How will Russia's invasion of Ukraine affect global food security?

Joseph Glauber and David Laborde

Originally published February 24, 2022.

The unfolding crisis in Ukraine has roiled commodity markets and threatens global food security. Ongoing fallout from the COVID-19 pandemic and other factors have already driven up food prices. Poor harvests in South America, strong global demand, and supply chain issues have reduced grain and oilseed inventories and driven prices to their highest levels since 2011–2013. Vegetable oil prices have also been at record levels, reflecting the short South American soybean crop, reduced palm oil supplies due to harvest problems in Malaysia, and sharply increased use of palm and soybean oil for biodiesel production. Prices of key energy-intensive inputs like fuel, fertilizer, and pesticides have also been at near-record levels.

Russia's invasion of Ukraine will further disrupt global markets, will have negative consequences for global grain supplies in the short term, and by disrupting natural gas and fertilizer markets, have negative impacts for producers as they enter a new planting season. This could push up already-high food price inflation, and have serious consequences for low-income net-food-importing countries, many of which have seen an increase in malnourishment rates over the past few years in the face of pandemic disruptions.

In this post, we discuss the global implications of the crisis; however, we should not forget the direct impacts of the war on local populations. Military operations could lead to the displacement of 1.5 million to 5 million people in Ukraine, leading to a major food crisis. As the State of Food Security and Nutrition and Global Report on Food Crisis reports point out, conflicts remain a key driver of food insecurity in the world.

The rising role of the Black Sea in global food security

Over the past 30 years, the Black Sea region has emerged as an important global supplier of grains and oilseeds, including vegetable oils (Figure 1). In the early 1990s, following the breakup of the former Soviet Union, the region was a net importer of grain. Today, Russia and Ukraine's exports account for about 12% of total calories traded in the world, and the two countries are among the top five global exporters for many important cereals and oilseeds, including wheat, barley, sunflower seeds, and maize. Ukraine is also an important source of sunflower seed oil, supplying about 50% of the global market.

Many importing countries depend heavily on these products from Ukraine and Russia. North Africa and the Middle East (MENA) import over 50% of their cereal needs and a large share of wheat and barley from Ukraine and Russia. Ukraine is an important supplier of maize for the European Union and China, as well as several North African markets including Egypt and Libya.

Likely short-term impacts

The crisis will likely have an immediate impact on grain exports from Ukraine and Russia. Most of the wheat and barley crops are harvested in the summer and exported during the fall. By February, most wheat, barley, and sunflower seed exports are largely completed. Ukraine maize exports typically remain heavy through the

Figure 1: Ukraine and Russia's share of global trade, 2018-2020

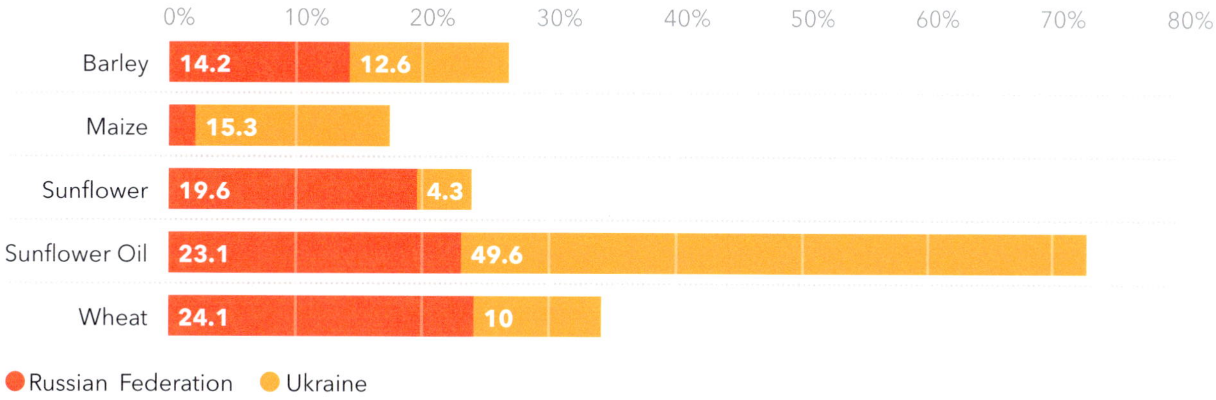

Note: Intra-EU trade excluded from computations.
Source: COMTRADE

spring into the early summer. Most grain flows out of Odessa and other western ports on the Black Sea, far from the occupied areas in the East, but disruptions are looking increasingly likely given Russia's latest military actions. Military operations could have short- and long-term consequences for the capacity to move Ukraine's crop production within and beyond its borders, especially if port facilities and railroads are damaged through terrestrial and aerial operations, or by cyber-attacks targeting various infrastructures and their management.

Looking forward to 2022 crops, the Luhansk and Donetsk Oblasts account for about 5% of Ukraine's barley production, 8% of wheat production, 9% of sunflower seed production, and a negligible share of maize production. The occupied areas of Luhansk and Donetsk are in their easternmost parts, bordering Russia. However, large areas of production are in other parts of Ukraine that directly border Russia and Belarus where Russian troops have also massed – between 25 to 30% of maize and sunflower seed production, 10 to 15% of barley production, and 20 to 25% of wheat production are in such oblasts. Spring barley will be planted in March, while maize planting typically begins in April. Winter wheat typically won't be planted until September.

Longer-term effects: Impacts on fertilizer and energy inputs

The Russian invasion of Ukraine will likely provoke countermeasures by the EU, United States, and other countries, which could have large impacts on Russian exports of natural gas and fertilizers. Russian exports of natural gas account for about 20% of global trade and Russia supplies about 40% of the EU's current imports. Sanctions could halt trade and drive up natural gas prices to even higher levels (Figure 2). While European importers could switch to other suppliers such as the United States, logistical issues – the US exports liquified natural gas – add costs, and these supplies would not provide significant relief, at least in the short term.

Natural gas is also an important feedstock for the production of nitrogenous fertilizers such as ammonia and urea. The impact on fertilizer prices would be further exacerbated by the fact that Russia is an important supplier of nitrogenous fertilizers and potash. Russia account for 15% of global trade in nitrogenous fertilizers and 17% of global potash fertilizer exports. Belarus, an ally of Russia and staging ground for the current invasion and already being targeted by some international sanctions, accounts for an additional 16% of global market share of potash exports. Dependency for some countries, including Ukraine, on the supply from these two countries is quite extreme (60% or more, see Figure 3).

Figure 2: **European and US natural gas evolution, US dollars per MMBTu**

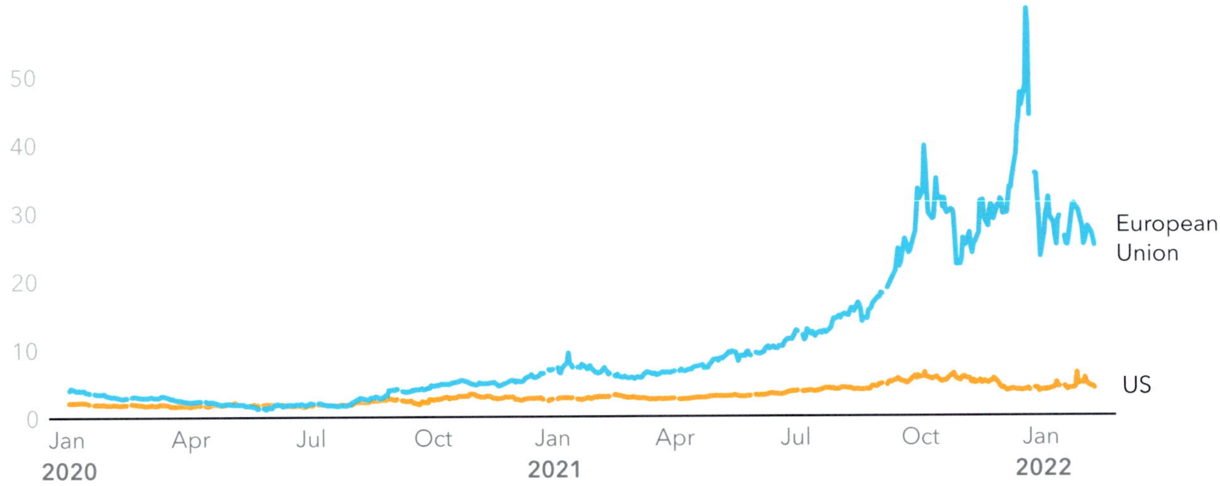

Note: US price based on Natural Gas, next month, contract. European price based on Dutch TTF contract. MMBTu = million British Thermal Units.

The global fertilizer market has been reeling from record high prices. Further shortages will have global implications, particularly in developing countries where price impacts could significantly reduce fertilizer use and result in poor local harvests at a time of reduced global stocks and record global prices.

Conclusion: Possible next steps

Russia's invasion of Ukraine poses serious risks to global food security that will require a range of responses from governments and international organizations. The situation remains very fluid and introduces considerable uncertainty into an already tight global marketplace. Of immediate concern is the vulnerability of net-food-importing countries highly dependent on Ukraine for supplies – such as the highly exposed countries in the MENA region. The global response to the Ukrainian situation should include a strong global food security dimension, and ensure that any sanctions placed on Russia not impact third parties reliant on Russian exports. To the extent possible, food and fertilizer exports should be allowed to continue unimpeded; failing that, mitigation packages should be provided to affected third-party countries. Placing sanctions on those sectors will only exacerbate world shortages and penalize mainly populations that are already food insecure.

With skyrocketing agricultural prices, there is a risk that some countries may seek to insulate domestic producers by restricting exports. Such moves should be avoided. As we saw in 2007/08 and 2010/11, these beggar-thy-neighbor policies can have harmful effects on importing countries, particularly the most vulnerable. In this context, both the sanctions aimed to enforce international laws and export restrictions used to protect domestic consumers have to be designed to protect global food security, and consequences for third parties need to be assessed carefully.

Finally, the reliance of many countries in the world on Russian energy products will trigger important policy debates. Calls for greater energy sovereignty and diversification will likely occur. However, policies that call for ramping up production of biofuels should be viewed with skepticism. Redirection of the current consumption of food crops (maize, wheat, oilseeds) to non-food uses around the world, from the EU to Indonesia, is already generating significant tensions in food and fertilizer markets. A holistic approach regarding food and energy security is critical in ensuring that food and nutrition outcomes remain a priority.

Figure 3: **Countries' dependency on nitrogenous and potassium fertilizers originating from Belarus and the Russian Federation, 2018-2020**

A. Percentage of nitrogenous fertilizer imports

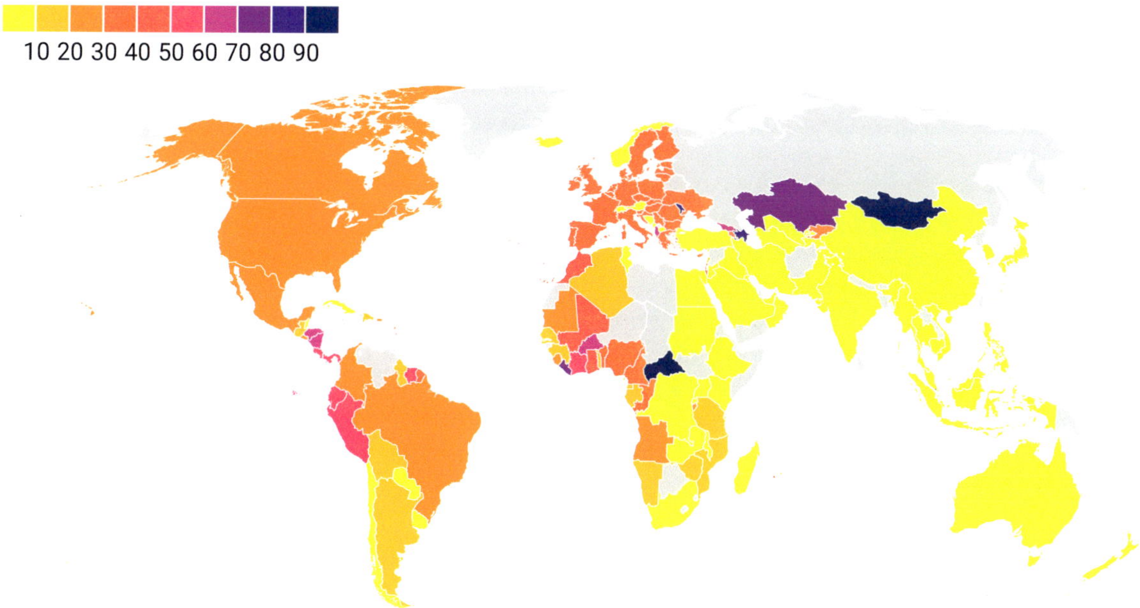

Source: COMTRADE.

B. Percentage of potassium fertilizer imports

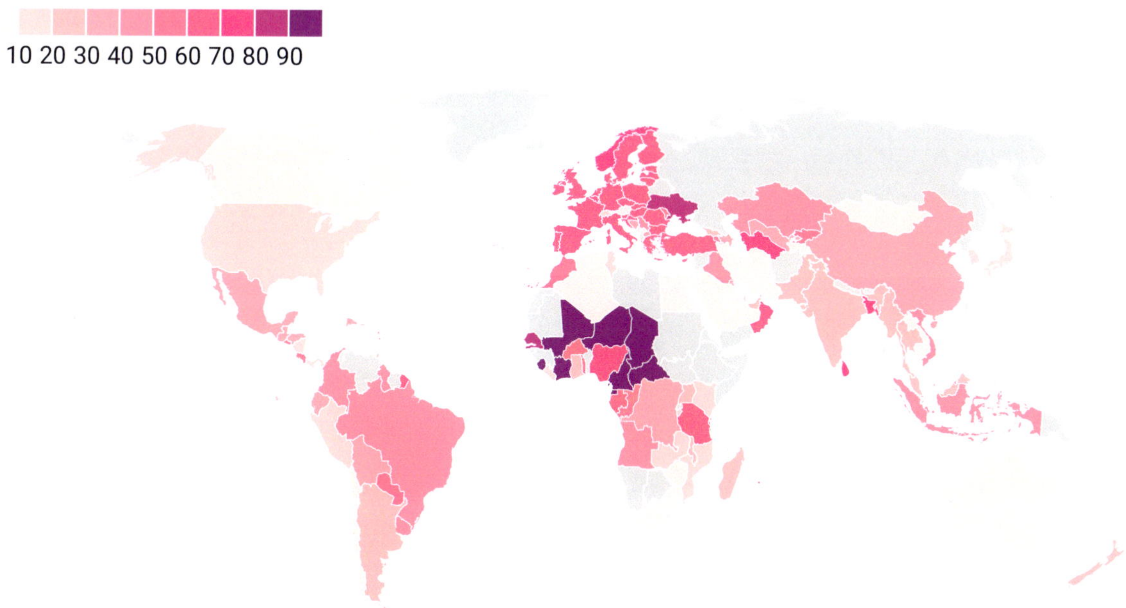

Source: COMTRADE.

Update

After peaking in May 2022, agricultural prices have generally fallen back to pre-war levels and price volatility has fallen as well. While some may conclude that the global food crisis is over, concerns remain. Despite the partial reopening of the Black Sea to shipping, the deal struck between Russia and Ukraine remains tenuous. Indeed, Russia's threat of pulling out of the agreement in late October pushed global wheat prices up by more than 5%. The current agreement runs through March 2023, and markets will be monitoring negotiations between the parties very carefully. In the meantime, fall wheat plantings in Ukraine are estimated to be down by more than 30%, reflecting the impact of the war itself, higher input costs, and depressed domestic prices due to the high costs of transporting grain and oilseeds to market.

As a result, the impact of the war on global crops will be felt again in 2023. Projections by the U.S. Department of Agriculture and others suggest that year-end stocks will remain tight this year with no significant rebuilding. Argentina's wheat and soybean crops are estimated to be down due to drought; and despite a record harvest in Australia, heavy rains in the Southeast may have affected quality. There has also been some increase in rice prices over the past six months, reflecting higher input costs (and lower margins). But the real concern remains the war itself. As long as the war continues, Ukraine's agriculture (accounting for 6% of global agricultural exports on a caloric basis) will continue to operate at a limited level. That export deficit will have to be made up elsewhere or food costs will remain high.

2 A food crisis was brewing even before the Ukraine war – but taking these three steps could help the most vulnerable

Derek Headey and Kalle Hirvonen

Originally published by IFPRI March 25, 2022.

Russia's invasion of Ukraine has led to the disruption, by sanctions or war, of two of the world's largest grain exporters. This means 2022 is shaping up to be a very difficult year for the global food system.

Yet there were concerns that this system was creaking at the seams as far back as 2007. At that time, there were steeply rising food prices driven by rising oil prices, explosive demand for corn-based biofuels, high shipping costs, financial market speculation, low grain reserves, severe weather disruptions in some major grain producers, and a swath of nervy trade policies leading to further shocks that worsened the problem.

The World Food Program's director general described it as a "perfect storm." Prices spiked again in 2011/12 before gradually receding.

In retrospect, those storms might now appear temperate in comparison to that we face in 2022. Even before the current crisis unfolded, food, fertilizer, oil, and shipping costs were rising steeply.

The FAO cereal price index shows prices hit their 2008 level in 2021, and since the invasion they have exploded. Between 2019 and March 2022, cereal prices increased by 48%, fuel prices by 86%, and fertilizer prices by 35% (Figure 1).

Here are three factors that we think make the situation in 2022 much worse, and three measures that could help prevent a global food supply system collapse.

The poor are still recovering from the COVID-19 crisis

Back in early 2008, both the developed and the developing world had just experienced an unprecedented period of rapid economic growth and poverty reduction, in some cases after decades of stagnation. The global financial crisis of 2007–2008 only briefly halted growth in the developing world. Many governments and international institutions recognized the need to reinvest in agriculture, and found resources to do so.

But fast-forward to 2022, and the world has not yet recovered from the repercussions of the COVID-19 pandemic, the worst economic crisis since World War II. There are no truly rigorous estimates of COVID's impact on global poverty, but the World Bank has estimated that 2020 saw an extra 97 million people thrown into poverty.

Household surveys and economic models consistently find that the pandemic had the most severe economic impact on the urban poor, while agriculture and the rural economy remained remarkably resilient to lockdowns and other demand shocks. Higher food prices may even have had a poverty-reducing effect in rural areas. But the urban poor lost out and, after two years of turmoil, are being hit hard again, now by food price inflation.

Figure 1: Food, fuel, and fertilizer prices vs. low- and middle-income country GDP growth, 2000–2022

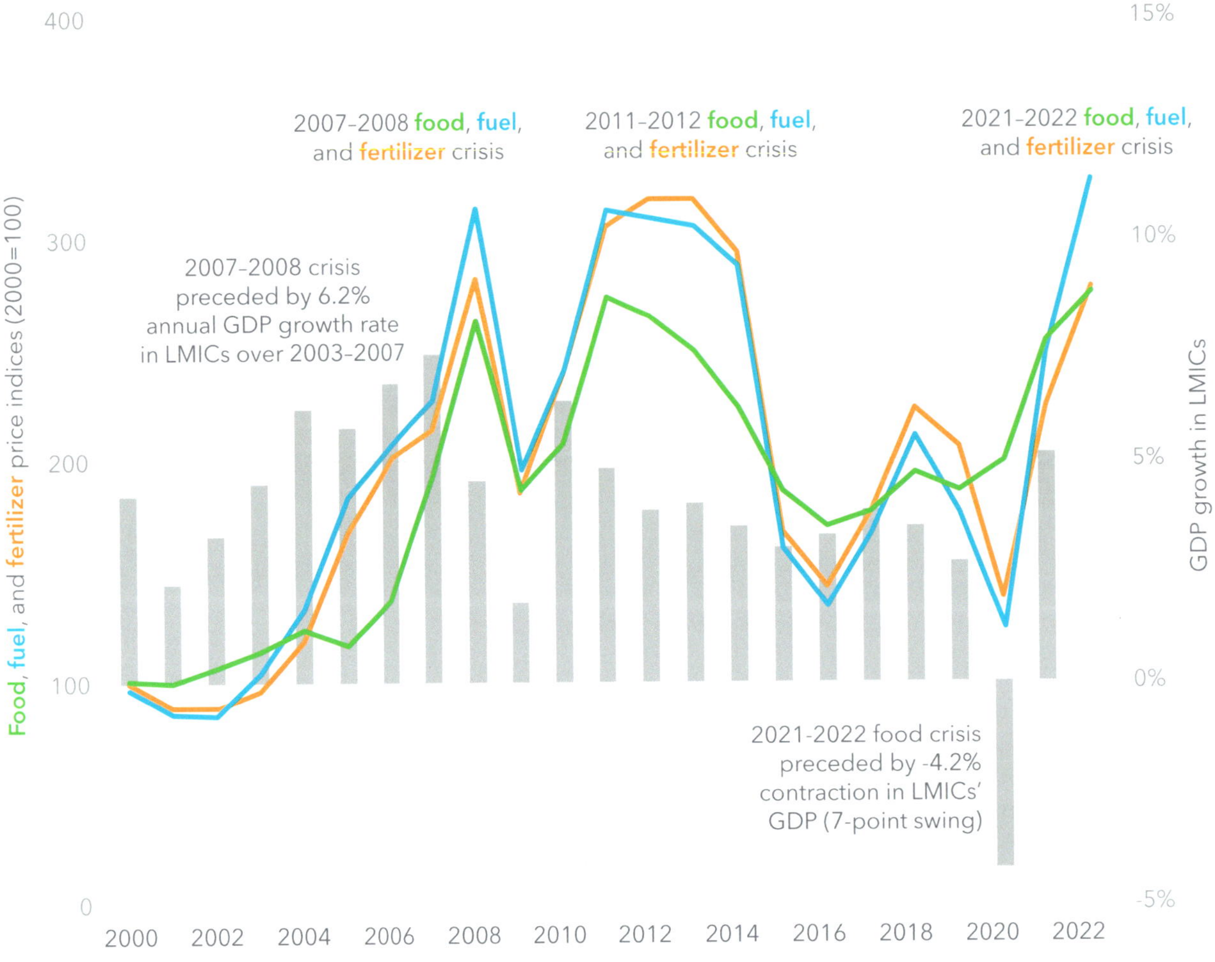

Source: FAO/IMF/World Bank

Cash-strapped governments have little room to maneuver

Both developed and developing economies are weaker today than in 2008. To their credit, governments in the developing world provided unprecedented protection for households and businesses and digital innovations for reaching the poor during the pandemic.

But as a result, many economies face large debt burdens relative to national income, as well as growing deficits, weak exchange rates, uncertain near-term economic growth prospects, and foreign investors and development partners that are also short of cash.

Africa is undoubtedly one of the most vulnerable regions. North Africa is a huge net importer of wheat, most of which comes from Russia and Ukraine, so it faces a particularly acute food crisis. Sub-Saharan Africa is predominantly rural, but its growing urban populations are relatively poor and more likely to consume imported grains.

Farmers in many parts of Africa are struggling to access fertilizers, even at inflated prices, due to shipping and foreign exchange problems. Exorbitantly high costs will erode farmers' profits and could reduce incentives to increase production, dampening the poverty-reduction benefits of higher food prices.

Countries already affected by conflict and climate change are exceptionally vulnerable. War-ravaged Yemen is heavily dependent on imported grains. Northern Ethiopia is one of the poorest regions on Earth, facing ongoing conflict and a humanitarian crisis. And Madagascar was slammed by successive tropical storms and cyclones in January and February, leaving its food system broken.

In Afghanistan, child mortality rates are soaring due to the collapse of the economy and basic health services. Myanmar's GDP shrunk by 18% after the military coup in February 2021, and food prices have since increased by 19%.

A crisis with no end yet in sight

The 2007/08 food crisis was relatively short lived, and the global food system responded swiftly with increased supply. But who can say with any confidence whether the effects of the Ukraine crisis on food, fuel, and fertilizer prices will end any time soon?

Russia and Ukraine account for more than 30% of global grain exports, Russia alone provides 13% of global fertilizer and 11% of oil exports, and Ukraine supplies half of the world's sunflower oil. In combination, this is a huge supply shock to the global food system, and a protracted war in Ukraine and the growing isolation of Russia's economy could keep food, fuel, and fertilizer prices high for years.

Three things that can be done

What can be done to buffer the worst impacts of this crisis?

- First, major grain producers must do everything they can to increase food supply: resolve logistical bottlenecks, release stocks, and resist the urge to impose food export restrictions. In particular, Southeast Asian countries must band together and avoid the crisis spreading to rice markets through trade restrictions.
- Second, in the short term the world needs oil-producing nations – often huge net importers of food – to increase fuel supplies to help bring down fuel, fertilizer, and shipping costs. This will benefit the entire global food system. Oil exporters can also step in to increase foreign assistance, especially for humanitarian aid. But in the long term we need to reduce global dependence on fossil fuels, including in agriculture.
- Third, governments, international institutions, and even the private sector must offer social protection via food or financial aid. The pandemic's effects hit the poor and vulnerable hard, and for much longer than expected.

Despite their sometimes dire fiscal circumstances, governments must again reach deep into their treasuries to reinvigorate this protection – and the international community must help them. There is simply no other alternative for averting a humanitarian disaster that will hit the developing world hard this year, and conceivably well into years to come.

This post first appeared on The Conversation.

3 The Russia-Ukraine war's impact on global food markets: A historical perspective

Joseph Glauber, David Laborde, and Johan Swinnen

Originally published April 6, 2023

The Russia-Ukraine war has focused global attention on the key economic roles those countries play as major exporters of agricultural commodities. Over the 2019–2021 period, they accounted for 12% of global agricultural trade on a kilocalorie basis, with a combined market share of 34% for wheat, 26% for barley, 17% for maize, and 75% for sunflower oil. The war has scrambled this picture, with Ukraine's exports falling dramatically, and Russia's falling, then recovering.

While the Black Sea region has historically been a major grain producer, its emergence as a major world exporting region is a relatively recent phenomenon. From the 1970s until the early 2000s – the decades immediately before and after the collapse of the Soviet Union – Russia and Ukraine were net grain importers. This post examines how and why Russia and Ukraine became such important factors in 21st century global food markets – and thus why the war poses a continuing threat to global food security.

Distortions in the Soviet Union and collapse in the 1990s transition

By the early 1970s, the Soviet Union had become a large net importer of wheat (Figure 1a) and maize (Figure 1b). This was the direct result of Soviet economic policies; specifically, production shifted away from grains after the government instituted a policy to stimulate meat and dairy production and consumption via large subsidies provided through massively distorted prices.

Grain imports continued in the years following the formal dissolution of the Soviet Union in 1991, but for different reasons. After 1991, Russia's agriculture sector, like the overall economy, collapsed. The region's abrupt transition from a planned to a market economy during the 1990s resulted in a severe decline in agricultural gross output. Dairy, livestock, and grain production all fell. Land reforms in Ukraine and Russia were introduced slowly as part of the economic transition. As collective farms were broken up, many small operations struggled, unable to take full advantage of scale efficiencies offered by mechanization and other technologies. Full private ownership of land with transfer rights evolved gradually; in Ukraine, a major land reform was introduced in 2002, but the moratorium on land sales transactions was only lifted in 2021. At the same time, capital constraints restricted fertilizer use and disruptions in supply chains made grain trade costly.

As a result of these problems, cereal area in Russia fell by over 25% in the 1990s, from almost 60 million hectares in 1992 to 43 million in 2001. In addition, grain yields in Russia and Ukraine declined strongly in the 1990s as fertilizer use fell by about two-thirds. As a result, both Ukraine and Russia continued to be net importers of cereals.

Productivity growth since 2000

Agricultural production in Russia and Ukraine began to turn around after 2000. By then, improvements in property rights and farm restructuring efforts in the 1990s began yielding increasing productivity. Around the

Figure 1: Net wheat and maize exports from the former Soviet Union, metric tons

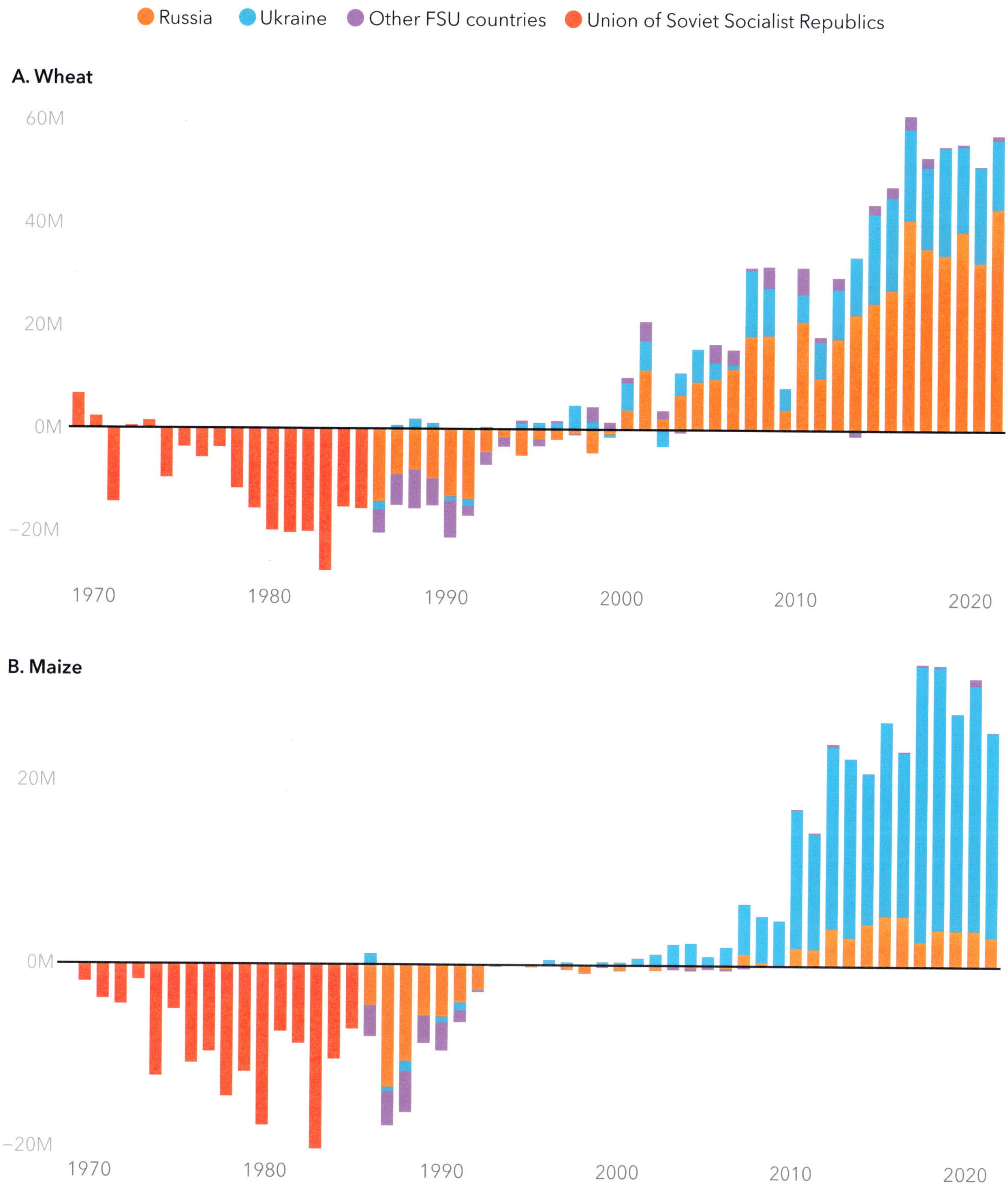

Note: July/June marketing year, 1970 = 1970/1971

Source: USDA/FAS PSD database, February 8, 2023

Figure 2: Former Soviet Union wheat and maize production, metric tons

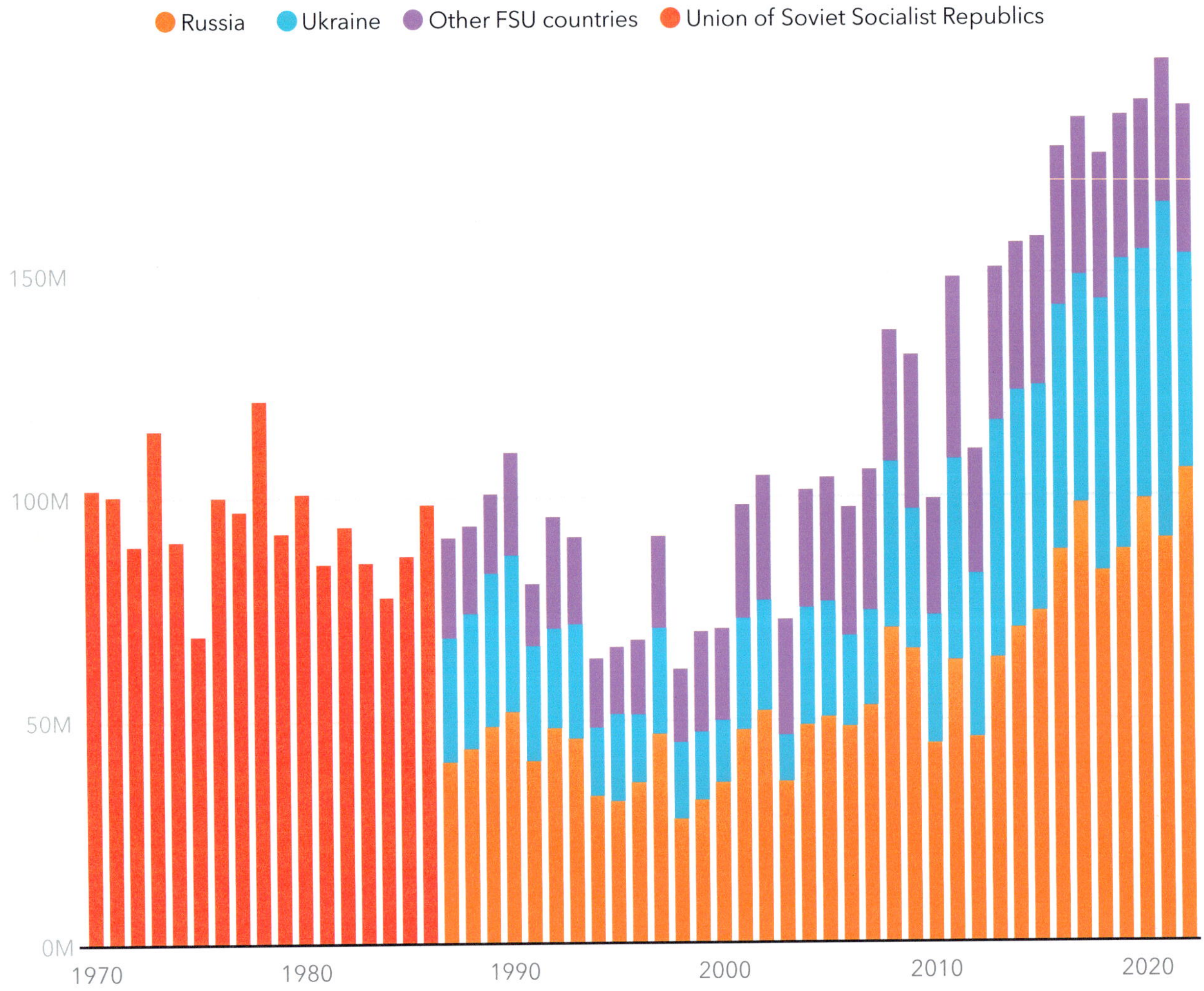

Source: USDA/FAS PSD database, February 8, 2023

same time, the Russian and Ukrainian economies recovered from the late 1990s financial crisis and investments increased in the agriculture sector.

Wheat yields for Russia and Ukraine rose steadily – by 1.4% and 1.8% annually, respectively – between 2002/02 and 2021/22. Maize yields showed even larger growth. Ukraine's maize yields have risen by 4.4% annually over the last 20 years, while Russian maize yields have risen by 5.5% per year.

Nevertheless, many problems remain. Cereal production in Russia and Ukraine has shown a relatively high degree of variability during the last two decades, and research suggests that yields of these crops in Ukraine and Russia have not reached their maximum potential. The relatively high vulnerability of crop yields to adverse weather conditions and the relatively low yields compared to European Union levels are also related to limited access to credit and finance and uncertainties in the regulatory framework of the agriculture sector, which create a barrier for input use (such as fertilizer and plant protection) and general investments.

Figure 3: Russian and Ukrainian crop yields, metric tons per harvested hectare

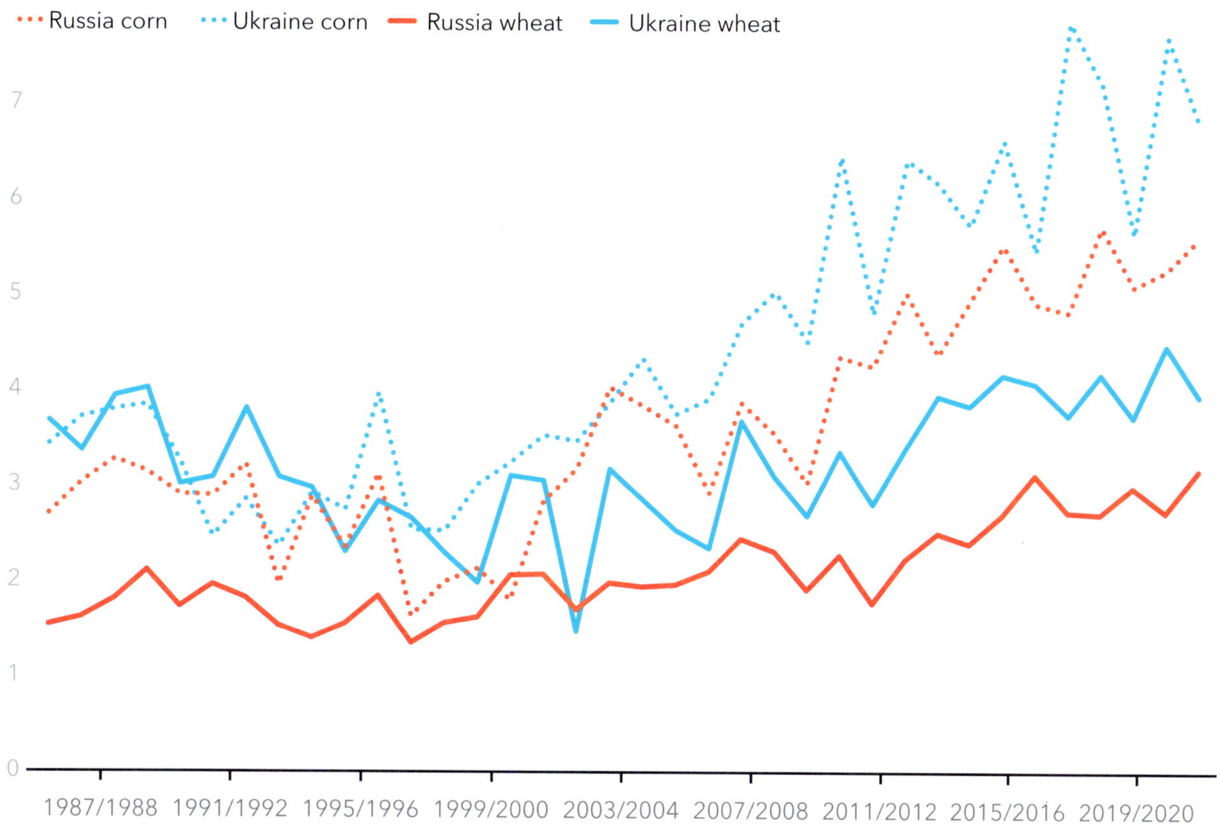

Source: USDA/FAS PSD database, February 8, 2023

Growth of exports and new markets

As production rebounded in the 2000s (Figure 2), grain exports increased dramatically and accelerated in the past decade. From 2012/13 to 2021/22, annual wheat exports increased by 10.0% annually in Russia and 11.3% in Ukraine, while the rest of the world's wheat exports increased by only 2.3% per year. Over the same 10-year period, Russian and Ukrainian maize exports increased by 7.6% and 6.5% per year, respectively.

What are the principal market destinations for Russian and Ukrainian crops? Given the proximity of the Black Sea, countries in North Africa and Western Asia have been the primary destinations for wheat exports (Figure 4). Those countries combined accounted for about half of Russian and Ukraine wheat exports over the period 2019-2021.

Yet market destinations have also changed and evolved with expanding production. Wheat exports to the rest of Asia have risen over the past 20 years and accounted for 13% of Russian wheat exports over the period 2019-2021. Countries such as Indonesia, the Philippines, and more recently Pakistan have become important buyers of Ukrainian wheat. Overall, Asian markets account for almost 40% of its wheat exports (2019-2021). African countries south of the Sahara are a small but growing market, accounting for almost 13% of Russia's exports and 8% of Ukraine's exports in 2019-2021.

Figure 4: Ukrainian and Russian wheat exports by destination, 2019-2021, metric tons

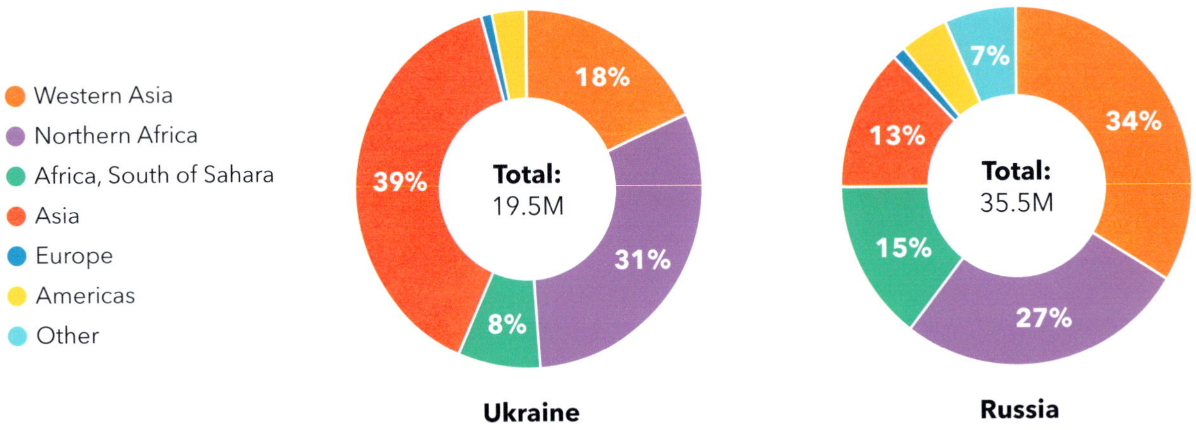

Source: COMTRADE

Figure 5: Ukrainian and Russian maize exports by destination, 2019-2021, metric tons

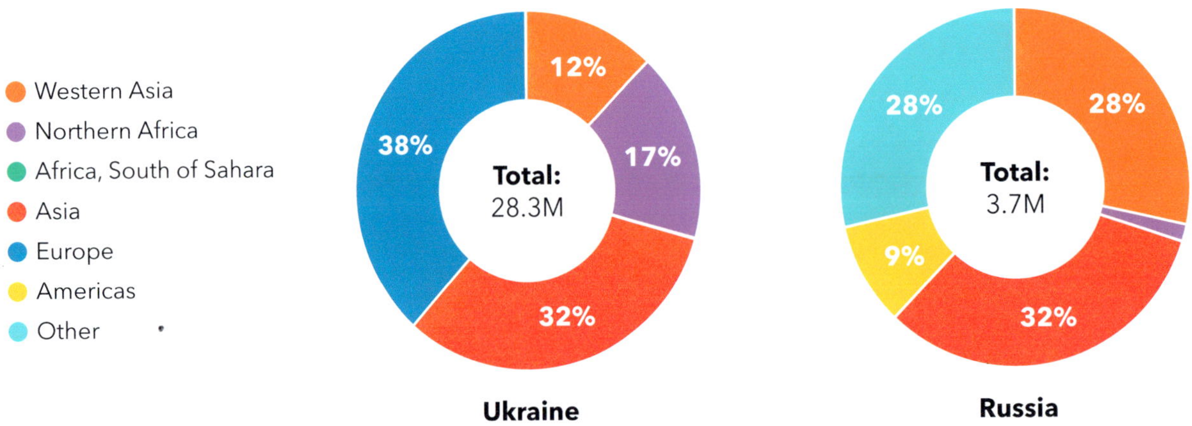

Source: COMTRADE

The market composition for Russian and Ukrainian maize is less similar than that of wheat exports (Figure 5). In recent years, both countries have shipped about almost one-third of their maize to Asian markets such as China. The remainder of Russia's maize exports go to markets in North Africa and Western Asia (largely Middle Eastern countries like Syria and Türkiye). North Africa and Western Asia are important markets for Ukraine as well (accounting for almost 30% of its exports over 2019 to 2021). The remainder of Ukraine's maize exports go largely to Europe (about 40%), whose feed manufacturers have a preference for non-GMO maize. Neither Russia nor Ukraine is a significant exporter of maize to African countries south of the Sahara.

Increasing reliance on Black Sea grains

The expanding production and exports of Russian and Ukrainian agricultural products over the past two decades has made them important players in global markets. This increased many countries' reliance on the region to meet key import needs (Figure 6) – and made their populations more vulnerable to the war's disruptions. For example, Ukraine and Russia accounted for 50% of Egypt's wheat imports in 2013; by 2021, they accounted for over 76% of its imports. Indonesia imported 8% of its wheat needs from the two countries in 2013 and almost 27% in 2021. Ukraine accounts for over 55% of the European Union's maize imports compared to less than 5% prior to the mid-2000s. China has become a large importer of maize as well, and Ukraine has been an important supplier, accounting for over 40% of China's maize imports over the past three years. At the height of its trade war with the United States, China imported over 80% of its maize needs from Ukraine.

Figure 6: Share of Russian Federation and Ukraine in imported calories

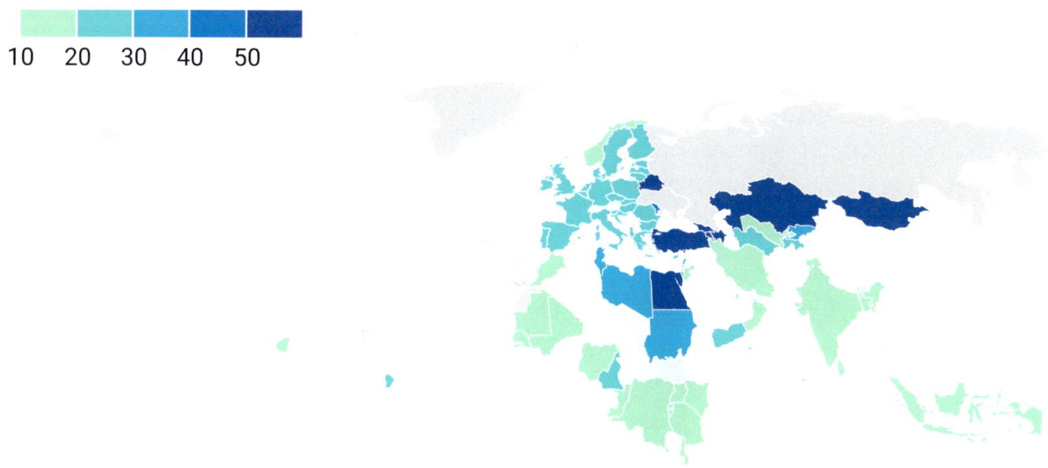

Note: EU countries are considered as one market.

The Russian invasion of Ukraine has sharply disrupted trade flows out of the Black Sea, as exports out of Ukraine's ports were blocked, forcing traders in grains and other agricultural products to seek alternative (and more costly) export routes. While evidence suggests that Russian grain exports have largely recovered, Ukraine's monthly export volumes continue to be well below levels of a year ago. This has caused many importing countries such as Egypt and Bangladesh to seek alternative suppliers.

Conclusions

The grain agreement between Ukraine and Russia has brought some hope that agricultural exports from the Black Sea region will return to more normal levels and that trade with the region will continue to grow. However, with the conflict ongoing, the agreement is at best tenuous. Production prospects in Ukraine for 2023 remain unpromising, as producers continue to face high costs of fuel and other inputs, combined with low output costs due to several factors, including a lack of sufficient storage and, despite the opening of trade corridors, high shipping costs that reduce local prices. The war's impacts on Russia's agriculture sector include sanctions imposed by Western countries that have raised input costs. How quickly the region rebounds – and when global market shifts and food security impacts settle down – will depend largely on how quickly the conflict comes to an end.

4 The Russia-Ukraine war is exacerbating international food price volatility

Brendan Rice, Manuel A. Hernández, Joseph Glauber, and Rob Vos

Originally published March 30, 2022.

The IFPRI Food Security Portal's Excessive Food Price Variability Early Warning System is showing excessive levels of price volatility in the four major food commodities: wheat, maize, rice, and soybeans, as well as cotton. Markets for hard and soft wheat and soybeans had already been more volatile than normal since late 2021, well ahead of Russia's invasion in Ukraine, which began on February 24, 2022. That conflict, coming on top of the ongoing effects of the COVID-19 pandemic, has already contributed to spiking food prices, with possible major consequences for global food security.

Rising price volatility poses a distinct threat, as it induces greater market uncertainty, which affects production decisions, and can spur speculative behavior. Both would fan further food price inflation. What is driving the current price volatility, and what are its implications for markets and food security?

The dangers of high price volatility

Tracking food price volatility provides a temperature check on global food markets. When prices fluctuate, it is more difficult for farmers to make decisions about what to produce and how; businesses are more reluctant to invest in food and agriculture; and ultimately consumption decisions are affected through higher prices and/or lesser availability, particularly in rural areas that depend heavily on agriculture and among low-income households who spend most of their income on food.

Volatility was high during the 2007/08 and 2010/11 global food price crises. Prices of staple foods rose quickly, followed by a period of high price variability that created major uncertainty in food markets. The excessive price variability tool was developed precisely to provide early warning of unusual price movements to market actors, including farmers, traders, investors, and policymakers.

The current situation compared to the early pandemic period

Food prices were already high before the Russia-Ukraine war. Poor harvests in South America, strong global demand, and pandemic-related supply chain issues had reduced grain and oilseed inventories and drove prices to their highest levels since 2011–2013. Prices of key energy-intensive inputs such as fertilizer were and continue to be at near-record levels. Then came the invasion, which is creating major market disruptions. Russia and Ukraine account for 30% of global wheat exports and supply millions of tons of wheat to food import-dependent developing countries in the Middle East and North Africa (MENA), South Asia, and sub-Saharan Africa. An ongoing war will certainly drive prices higher still and erode food security for hundreds of millions of people.

Looking back, most staple food prices also showed excessive volatility in the aftermath of the global food price and financial crises until 2012 (Figure 1). World market prices for basic staples were then calm for the most part until the recent rise in volatility.

Figure 1: Price volatility of major food commodities

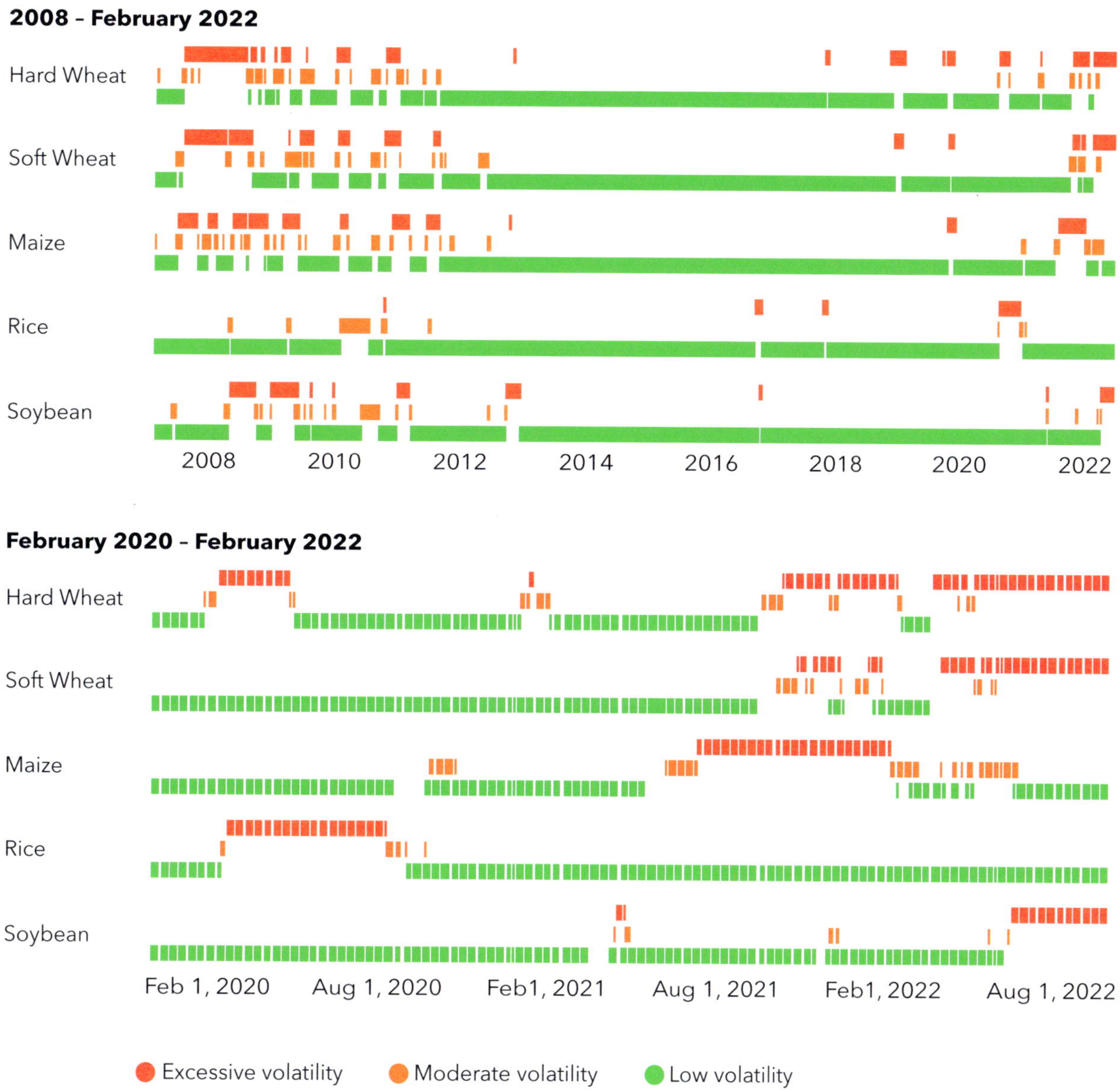

Source: Excessive Food Price Variability Early Warning System, Food Security Portal, facilitated by IFPRI.

Among the major staples, prices of rice and wheat became very volatile following the onset of COVID-19 lockdowns, driven in part by export restrictions imposed by some major producing countries. Russia and Kazakhstan, for example, imposed export bans on wheat, and Viet Nam imposed a short-lived ban on rice exports. In contrast to the 2007/08 global food price crisis, which led many countries to impose prolonged trade restrictions, most of these more limited recent measures were retracted by mid-2020 and staple food markets returned to lower, "normal" levels of volatility. This period of relative calm lasted only through the first part of 2021, as the deepening of pandemic-related supply chain disruptions mixed with strong demand driven by the global economic recovery.

Price volatility has continued to intensify since reaching excessive levels for wheat, maize, and soybeans in the final months of 2021. Markets for wheat and soybeans have remained volatile since, while that for maize returned temporarily to calmer waters. Following the invasion of Ukraine, however, maize prices again turned highly volatile, as did rice prices, so prices for all four staple foods are now in that state [as of March 2022]. Here is the situation for each:

- **Wheat:** Drought in many wheat-producing regions, including North America and the MENA region, along with strong import demand from China, has created concerns over adequacy of supplies, while shipping disruptions and grain export policy in Russia pre-war added to the uncertainty in global wheat markets. The invasion has exacerbated supply concerns, intensifying price volatility.

- **Maize:** Conflict-related supply concerns have also stoked price volatility in global maize markets. Ukraine is the world's fourth-largest maize exporter, and the war is putting the next harvest and the country's capacity to export in jeopardy.

- **Soybeans:** Drought in Brazil and Paraguay sharply lowered production prospects for soybeans, driving price volatility starting late in 2021. Market dynamics of other major oilseeds have also contributed to soybean price volatility. Increasing concerns over palm oil supply – particularly due to damage caused by typhoon-related flooding in Malaysia – have led to increased demand and higher prices for soybean oil, a palm oil substitute for food and fuel. Similarly, the conflict in the Black Sea region, an area that accounts for well over half of the world's supply of sunflower seed oil, has sent buyers scrambling for alternatives – also contributing to the excessive price volatility seen in soybean markets.

- **Rice:** The war has increased feed demand for rice on the heels of supply concerns in maize and wheat, pushing rice prices into excessive volatility territory.

The importance of monitoring food price volatility in the near term

Some conditions that contributed to price volatility following the initial COVID-19 lockdowns remain relevant now. Supply chain disruptions continue to be a factor, and tracking export restrictions is again important since such policies could lead food prices to spiral even higher.

One difference between the two periods is the scale of the disruptions in staple food markets. While the period of initial pandemic lockdowns saw some isolated volatility, the Russia-Ukraine war is affecting all major food staples. In addition, fertilizer prices are a greater factor now than at the beginning of the pandemic. Already at extremely high levels before the war began, fertilizer prices could continue rising as potash exports from major producer Belarus are cut off and Russia, another important fertilizer producer, is considering an export ban. High prices for natural gas, a feedstock for nitrogen-based fertilizers such as urea and ammonia, have boosted fertilizer prices as well.

Higher fertilizer prices could depress production, leading to less grain on the market in 2022 and putting further upward pressure on already-high food prices. With the conflict still unfolding and outcomes highly uncertain, monitoring food price volatility is more relevant now than at any time since the food price crises of 2007/08 and 2010/11 – with the food security of millions at stake.

5 No end in sight yet for the global food price crisis

Joseph Glauber, Manuel Hernández, David Laborde, Will Martin, Brendan Rice, and Rob Vos

Originally published September 27, 2022.

After the sharp rise in international prices of wheat and other staple foods in the wake of Russia's February invasion of Ukraine, since May prices have fallen back to pre-war levels. Has the global food price crisis now come to an end? Unfortunately, such a conclusion is premature. Domestic food prices for consumers continue to rise in most countries. Meanwhile, ongoing uncertainties – not the least of which is the continuing war – augur continued turmoil in global food markets. Global food security remains at high risk; hundreds of millions of people already face acute food insecurity and their numbers are rising, according to the Global Report on Food Crises. In this blog post, we try to disentangle the main factors in play.

Given the importance of both Russia and Ukraine as suppliers in global markets for wheat, maize, sunflower seeds and oil, and Russia's importance in international fertilizer and energy markets, the war provoked a surge in food and energy prices worldwide. The Food and Agriculture Organization (FAO)'s international food price index climbed to its all-time high in mid-May (Figure 1). At that point, wheat prices, for instance, were up 32% from their level at the start of the war. Then prices fell, with those of agricultural commodities, including wheat,

Figure 1: FAO monthly nominal food price index, December 2019–August 2022

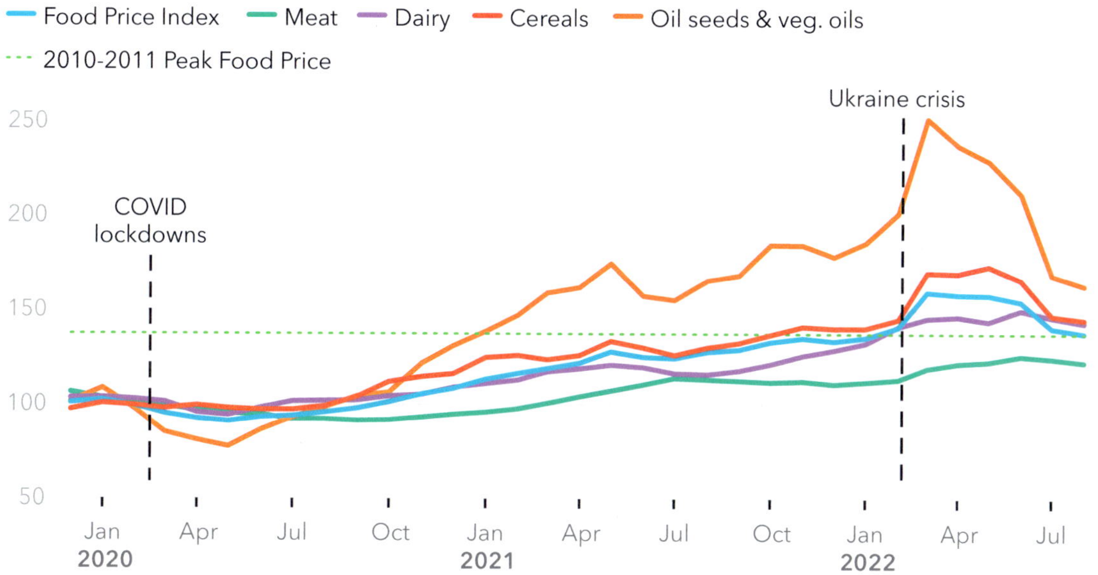

Source: FAO, Food Prices Index, accessed September 12, 2022.

returning to pre-war levels. However, prices remain high by historical standards, reflecting impacts of supply disruptions and the surge in global demand for commodities during the recovery from the COVID-19-induced recession in 2021.

High commodity prices alone need not pose risks to food security. Rather, it is the continuing volatility in tight markets and the degree of transmission to domestic food prices – that is, what consumers are actually paying – that we should be concerned about. Food price inflation is still up in most countries. In January 2022, consumer food prices in many countries had already increased 5% to 10% over the previous year. By July, consumers in a majority of both high- and low-income countries saw purchasing power go deeper into the red as they faced food price increases of between 10% and 30% – or even more in some contexts (countries marked in purple in Figure 2).

Figure 2: Consumer price inflation, October 2022

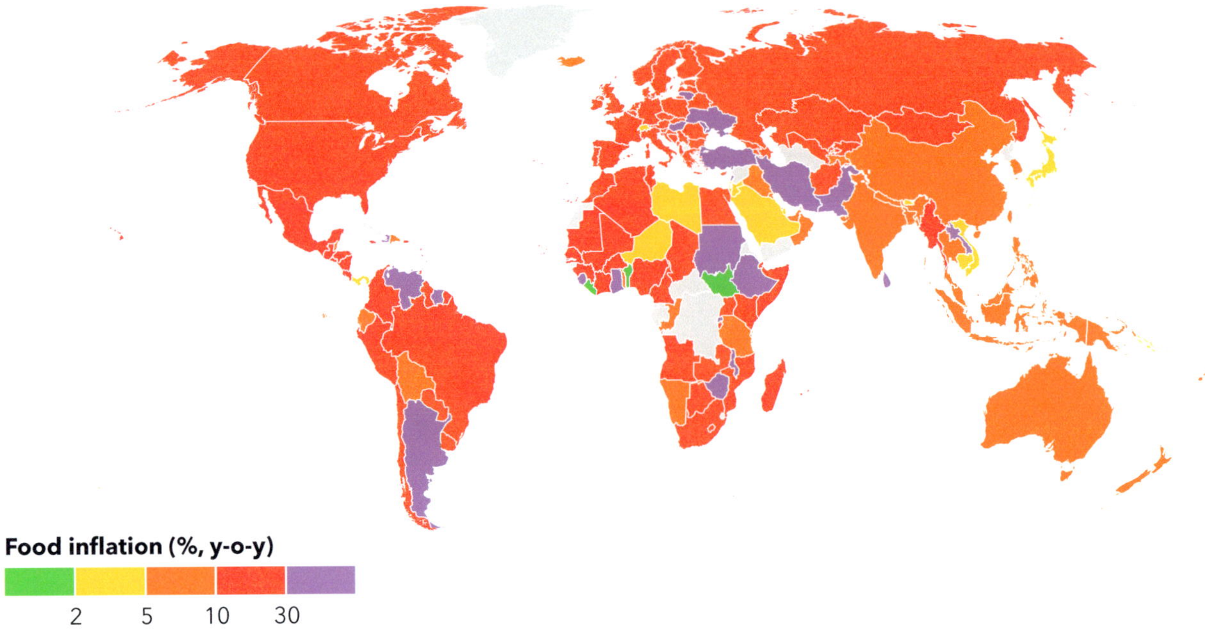

Source: Food Security Portal based on data from IMF and Trading economics

Understanding the price surge

The Ukraine shock came when most agricultural commodity markets were already very tight, with wheat and maize stock-to-use ratios at levels comparable to the lows of the 2007/08 global food price crisis. Market tightness was also reflected in high price volatility, which was already high in the months leading up to the war (see IFPRI's Excessive Price Volatility Early Warning System). As explained in our previous post on food price trends, during 2021, commodity prices rose with the uneven recovery from the COVID-19 crisis. Global demand surged, accompanied by supply disruptions related to transport and logistics bottlenecks – pushing up food, energy, and fertilizer prices as well as freight costs.

Russia's invasion of Ukraine further fueled food and fertilizer price inflation as it disrupted important supplies of wheat, maize, oil seeds, and fertilizer. Russia, meanwhile, is also a major exporter of fossil fuels, and war-related sanctions imposed on those exports pushed up the cost of nitrogen-based fertilizer production, which uses natural gas as a major feedstock and energy source. Fertilizer prices rose sharply.

In response to the crisis, many staple-food-producing countries introduced bans and other restrictions on exports, especially of wheat and vegetable oils, in efforts to keep food prices down for their own consumers, but – by reducing global supplies – pushing up prices for everybody else. By the end of April, 17% of globally traded staple foods (expressed in calories) were subject to such restrictions, according to IFPRI's Food and Fertilizer Trade Restrictions Tracker.

Freight costs had surged during 2021 but dropped dramatically toward the end of the year, easing global supply chain disruptions. Then they surged again as the war triggered new supply disruptions and uncertainties. The Baltic Dry Index, a key indicator of freight cost for food and fertilizer trade, roughly doubled between February and May 2022 (Figure 3).

Some observers are blaming market concentration and speculation as a factor driving up prices in global food (and fertilizer) markets. However, while many grain companies saw profits soar in the first half of 2022, it is less obvious that these companies have been driving the surge in prices.

Figure 3: Baltic Dry Index of international bulk shipping costs, December 2019–August 2022

Note: The Baltic Dry Index (BDI) is an index of average prices paid for the transport of dry bulk materials across more than 20 routes.
Source: The Baltic Exchange. Accessed September 12, 2022

What goes up….

The more recent price declines are most notable for wheat and maize, while rice prices have remained relatively flat, and international prices for meat and dairy are still on a moderate upward trend. While prices of cereals have come down in recent months, they are still up from a year ago and remain well above pre-COVID-19 levels.

The recent decline in international prices can be explained by several factors:

- Global growth has slowed as the recovery from the COVID-19 recession has faded and new lockdowns have slowed China's economic growth. Commodity prices, including those for food and energy, eased with weaker global demand.
- Seasonal factors are also playing a role, with harvests for winter wheat and maize taking place during June–August, expanding supplies.
- The stronger US dollar has softened dollar prices of internationally traded commodities.
- A number of countries lifted or reduced export restrictions. Between May and August, the share of global cereal and vegetable oils trade subject to restrictions dropped from 17% to 7% of trade.
- Freight costs dropped with weaker global demand, resuming the downward trend that began shortly before the Ukraine crisis.
- The Black Sea Grain Initiative agreements, reached in July, allowed shipments to resume from three of seven major ports in Ukraine, lifting the country's monthly exports of wheat, maize, and sunflower seeds to about 3 million metric tons per month. While encouraging, these levels are still less than 50% of the volumes typically shipped out of the Black Sea at this time of year. Thus, the actual impact on global market prices remains limited, as Ukraine would need to export at least 7 million tons per month to free up enough space in silos and ships to export the incoming harvests (arriving in August for wheat and in October for maize). On the other hand, however, Russian wheat exports appear less hampered by sanctions than originally feared.

But we're not out of the woods yet. International food prices remain high by historical standards, markets remain tight, and high price volatility continues – especially for wheat and maize.

At the same time, the war continues with no end in sight. There are signs already that Ukrainian farmers are likely to plant far less wheat this fall. The war and sanctions have kept fertilizer prices volatile, and fertilizer costs have risen faster than agricultural output prices. In August, energy prices rose as Russia cut off supplies of natural gas through pipelines to Central and Western Europe – also leading to a halt to nitrogen-based fertilizer production by several European producers, further pushing up fertilizer costs.

Another factor is poor weather conditions. Hot and dry conditions are affecting early development of wheat in Argentina and maize harvests in the United States, China, and Europe (AMIS, September 2022). In addition, rice prices increased during August and September, driven in part by extreme weather in some major rice producing countries, including catastrophic floods in Pakistan, the world's fourth largest exporter of rice. (High fertilizer prices and India's recent restrictions on rice exports have also contributed.)

Amid this uncertainty, the risk of new price shocks remains high. Further problems with Ukraine production and shipments or any other geopolitical or weather shocks could trigger price surges. According to the World Bank's April 2022 Commodity Markets Outlook, the war has altered global patterns of trade, production, and consumption of commodities in ways that will keep prices at historically high levels through at least the end of 2024, exacerbating inflation and food insecurity.

Food security remains at risk

These threats to food security remain acute in many countries. Food price inflation has accelerated across the globe. Fertilizer prices are rising faster than farm output prices, eroding farmer incomes. Surging domestic food and energy prices are drivers of general consumer price inflation around the world, as a recent IMF assessment points out. As a result of these trends, domestic food prices have not followed the downturn in international prices. Relative to the world market price, the average domestic price of wheat, for instance, rose by almost 20% between April–May and July.

These pressures are disproportionately affecting the poor, who spend most of their incomes on food, energy, and other basic needs. The poor in food-importing low-income countries tend to be hit hardest. Many of those countries are facing a perfect storm of overlapping crises and shocks that have eroded their already weak capacity to insulate their vulnerable populations from the worst impacts.

First, the global COVID-19 recession reduced livelihoods in general, and policy responses during 2020 eroded both fiscal space and foreign exchange. To fill fiscal and balance-of-payments gaps, countries ran up their external debt; the resulting overhang is now impeding further fiscal responses and import capacity that could mitigate the impacts of rising prices. The recent shocks have further weakened affected countries' currencies, exacerbating the domestic cost of the international price shocks.

Figure 4: Rise in cost of basic food basket in 45 food crisis countries

% change in cost of food basket (Apr–Jun 2022 vs. 5-year average for that period)

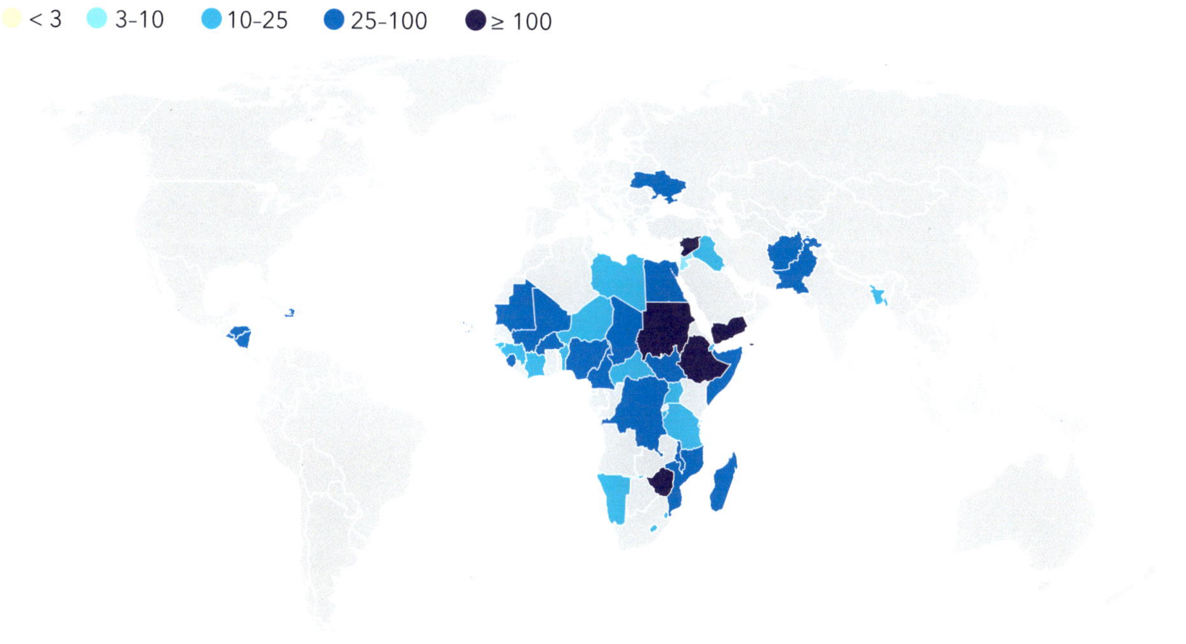

Note: The price of a typical food basket is based on meeting a population's energy and nutrition requirements. Its size and composition varies according to local preferences, demographic profile, activity levels, climatic conditions, etc. Data for Bangladesh refer to Cox's Bazar.

Source: WFP Global Market Monitor, August 2022; and FSIN, Global Report on Food Crises 2022 (mid-year update)

As the recent update of the Global Report on Food Crises points out, countries that were already facing protracted food crises before COVID-19 and the Russia-Ukraine war have been among the hardest hit by recent problems (Figure 4).

Populations in all 45 countries and territories already in food crisis before the war started had seen the cost of a basic food basket increase by at least 10% by April–June 2022 above the five-year average cost for that period). People in Sudan, South Sudan, Syria, Zimbabwe, Ethiopia, Yemen, Sierra Leone, Haiti, and Honduras faced a food cost increase of over 75%.

Most food crisis countries have high food import dependence and weak capacity to insulate their populations against imported food inflation due to low foreign currency reserves and/or depreciating national currency. Thus, they lack the capacity to expand social protection and other means to support vulnerable populations. In addition, many are witnessing high general price inflation, further eroding purchasing power. Protracted civil strife and weather shocks have compounded the economic woes in many of these countries, such that the total number of people facing crisis-level or worse acute food insecurity increased from 155 million in 2020 to 193 million in 2021 and to 205 million–223 million by mid-2022, according to the Global Report.

The global food crisis clearly is not over yet. The most vulnerable people continue to be hurt by sharply increasing food costs, seeing little relief even as global prices have fallen. While immediate humanitarian assistance will be necessary to avert more widespread famine, governments and international organizations should also focus on promoting major investments in more resilient and shock-proof food systems. This will require unprecedented international cooperation to address the consequences of unstable global commodity markets and climate-change-driven extreme weather.

6 The impact of the Ukraine crisis on the global vegetable oil market

Joseph Glauber, David Laborde, and Abdullah Mamun

Originally published May 3, 2022.

The war in Ukraine has pushed prices of agricultural products to historically high levels, and concerns about global food security occupy headlines and world leaders' minds, as demonstrated by recent International Monetary Fund and World Bank meetings. So far, much of the attention has focused on grains, particularly wheat – because of its importance in diets, and the predicament of countries where wheat accounts for a large share of calories consumed, is largely imported, and is dominated by supplies from the Black Sea.

Here, we focus on another important emerging food security issue: The war's impact on vegetable oils. The Black Sea countries are large exporters of sunflower oil, and the crisis has pushed vegetable oil prices significantly higher, and also triggered trade policy responses around the world that further restrict supplies and raise prices.

This post examines the unfolding impacts on the global market for vegetable oils, including additional factors affecting prices such as biofuel policies and export restrictions in the Black Sea and elsewhere.

Vegetable oils are a key item in diets around the world and an essential source of fats, accounting for about 10% of daily caloric food supply (300 kcal per day per person), making them the second-most important food group after cereals. Vegetable oils are also a nutritional source of omega-3 and omega-6 fatty acids and vitamins E and K. Vegetable oils are, of course, an essential cooking item, particularly for poor consumers unable to shift to more expensive butter or other animal fat-based products.

As with a number of other commodities, prices for many vegetable oils were at very high levels prior to the invasion in February. Since then, vegetable oil prices have risen by almost 30% on average (see Figure 1).

Sunflower oil has been most directly affected with an increase of more than 40% since the day of the invasion. It accounts for about 13% of vegetable oils traded in global markets, and Ukraine and Russia account for about 50% and 25%, respectively, of the sunflower oil traded in the world. Since vegetable oils require little or no processing, high prices have already been passed through to consumers, and some retail shortages have been reported.

An overview of the vegetable oil market

Vegetable oils are heavily traded, with imports representing about 40% of global consumption, especially when compared to other commodities, for example, less than 20% for grains. Three-quarters of countries (Figure 2) depend on world markets for at least half of their domestic needs.

As a result, most countries are more exposed to vegetable oil export shocks compared to cereals, which are less traded and in which many countries are relatively self-sufficient.

Together, palm oil (58%), soybean oil (14%), sunflower oil (13%), and rapeseed (canola) oil (7%) account for

Figure 1: Daily vegetable oil prices, current US dollars

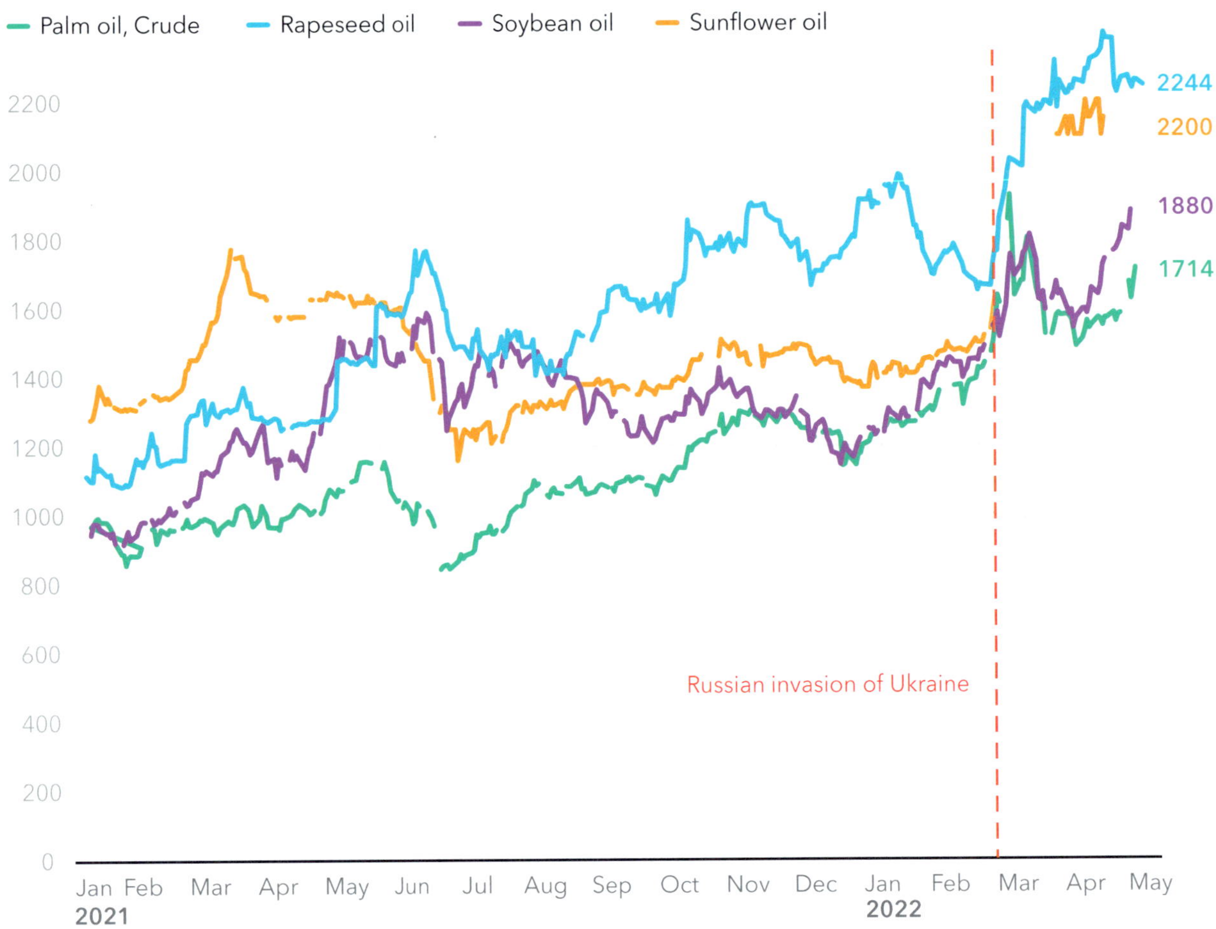

Note: Soybean oil: CBOT, Palm Oil, crude: Malaysian Board daily price, Rapeseed Oil: Rotterdam spot prices, Sunflower oil: India CIF Mumbai price
Source: MPOB, Bloomberg, CBOT

92% of vegetable oils traded in world markets on average between 2019 and 2021. The remaining 8% includes diverse, locally important oils such as olive oil, cottonseed oil, peanut oil, safflower oil, and palm kernel oil.

Indonesia and Malaysia accounted for 92% of total palm oil exports over the 2019–2021 period. Major soybean oil exporters include Argentina (46% of global exports over the period), Brazil (10%), the European Union (8%), and the United States (8%). Canada is the major exporter of rapeseed (canola) oil, accounting for 58% of rapeseed oil exports, but Russia (13%), the European Union (8%), and Belarus (7%) are also important exporters.

Prior to the Russian invasion of Ukraine, global vegetable oil supplies had tightened due to a variety of factors. Drought in South America has caused a sharp reduction in soybean yields, particularly for its largest producer, Brazil. As a result, soybean exports for Argentina, Brazil, and Paraguay are forecast to decline by 5% from last year's levels and vegetable oil exports are estimated to be largely unchanged. Malaysian palm production declined due to the impacts of Typhoon Rai in December 2021 and continuing acute labor shortages and

Figure 2: Share of imported vegetable oils in total consumption

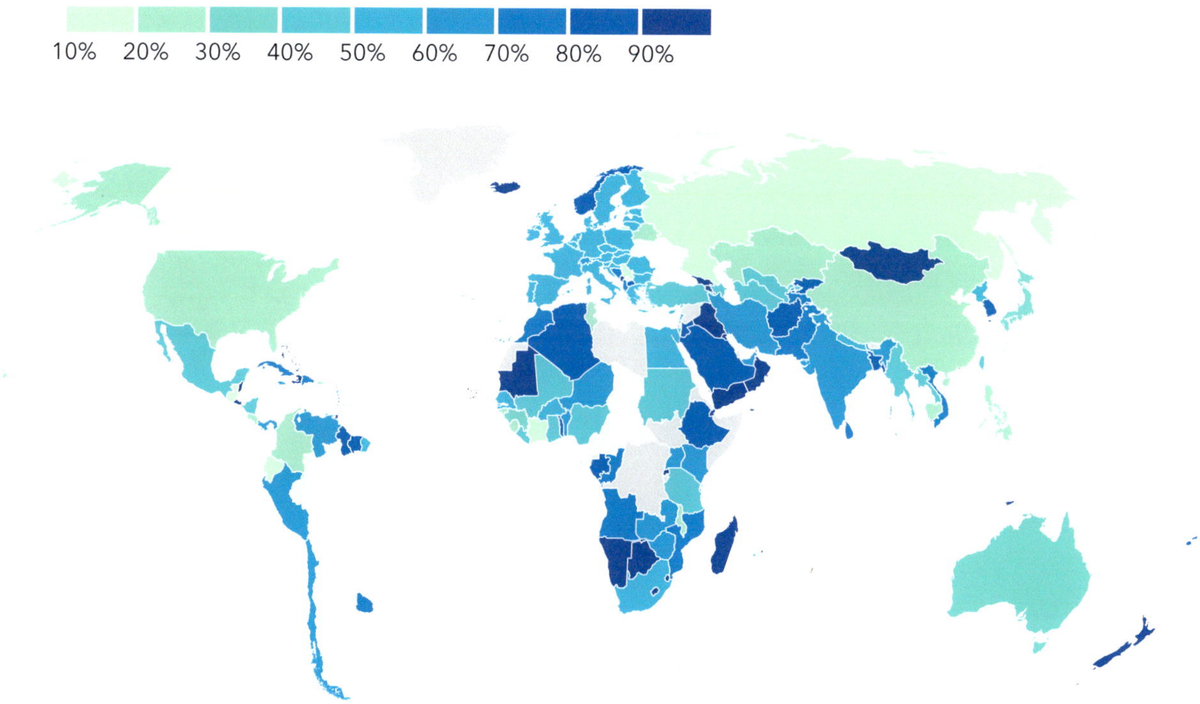

Note: European Union considered as an integrated market, only extra-EU trade considered.
Source: COMTRADE, FAOSTAT

other issues the industry has faced since 2019, amplified by the COVID-19 restrictions on worker mobility. Drought-impacted Canadian rapeseed production for 2021/22 declined 35% from the previous year, despite an 8% increase in planted area. As a result, Canada's rapeseed exports are estimated to fall by 50%, and rapeseed oil exports to be down year-over-year by 20%.

Then came the invasion. For vegetable oils, the most immediate global disruptions came in exports of sunflower oil. Important markets include India, the EU, and China, which together account for over 75% of total Ukraine exports. Port closures have prevented roughly half of the sunflower oil that typically would have been exported from last year's harvest from leaving in the country; the only alternative is more costly routes to the west through Romania or Poland. Processing facilities remain disrupted, and much of Ukraine's sunflower farming area lies to the east of the Dnipro (Dnieper) River, where much of the fighting is now. Russian exports are currently limited under the joint effects of maritime trade disruption in the Black Sea, the implementation of an export quota by Russian authorities, and the effects of sanctions on business transactions.

Vegetable oils vary in flavor and uses for different foods, but are largely substitutable for many purposes. As a result, vegetable oil prices tend to move together (Figure 1). Nonetheless, substitutions can create headaches for food manufacturers because of allergy issues and national labeling requirements. For example, France recently announced that food manufacturers could replace sunflower oil in their products but must give consumers a basic indication of any ingredient change within two months. These adjustments will create production bottlenecks and some shortages, while increasing production costs and at the end, increasing prices for consumers.

Figure 3: Share of imported vegetable oils in total consumption (%)

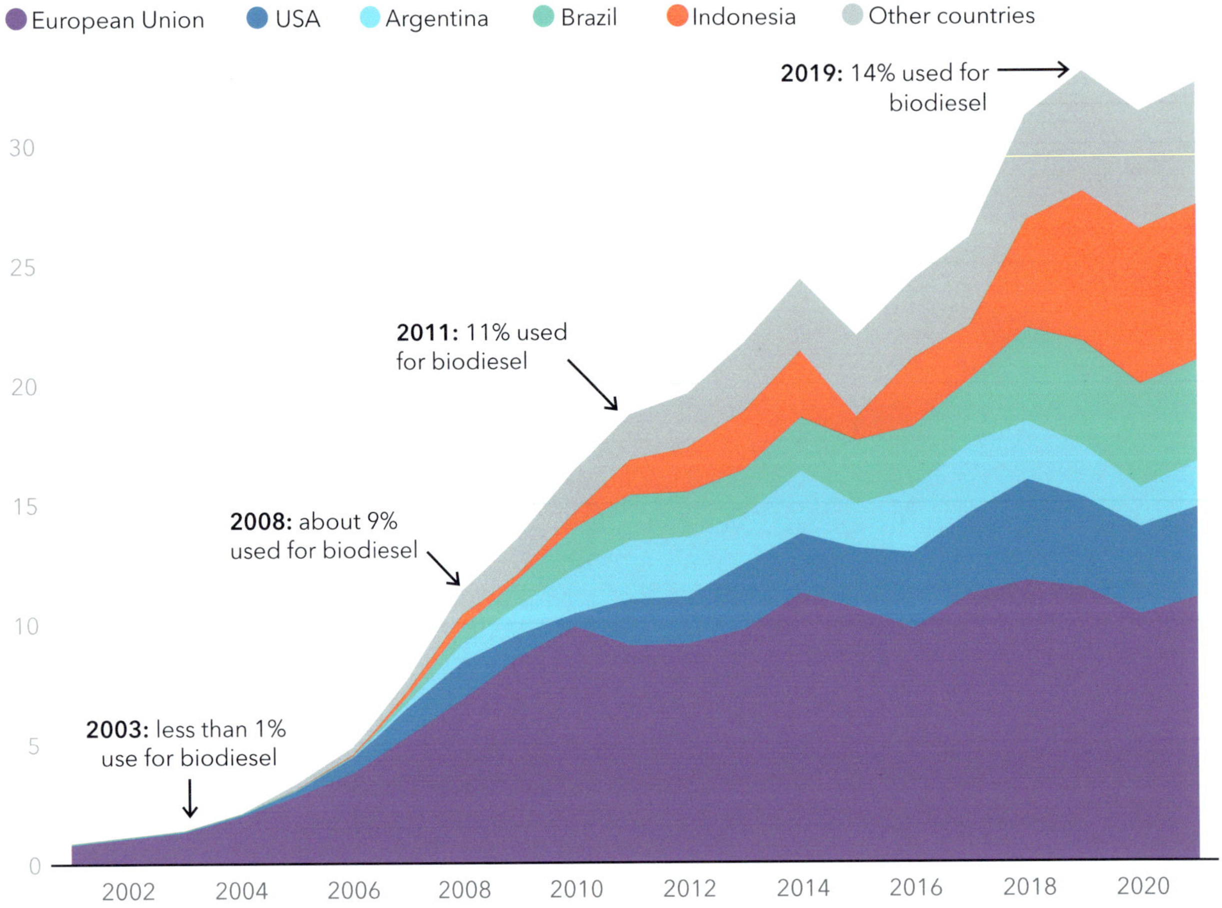

Source: FAO-OECD

Biodiesel production exacerbates supply shortages

Another issue affecting global vegetable oil supplies and prices is the recent growth of biodiesel capacity. Driven largely by regulations that mandate their blending in fuel supplies and subsidies encouraging their use, production of biodiesel fuels has grown over the past 20 years from less than 1% of total vegetable oil use in 2003 to almost 15% today (Figure 3). Major programs include the United States, Argentina, and Brazil (primarily soybean oil), Indonesia and Malaysia (palm oil), and the EU and Canada (primarily rapeseed oil and palm oil).

Biodiesel mandates are putting additional pressure on global vegetable oil prices. Because of mandates, demand for vegetable oils as a feedstock for biodiesel production is fairly unresponsive to changes in prices. As a result, price volatility tends to increase when supplies are short.

Export restrictions further threaten vegetable oil supplies

Amid the current crisis, trade restrictions imposed by major vegetable oil exporters have increased price volatility and with it, the vulnerability of food-insecure households around the world. Calculations based on IFPRI's

Export Restrictions Tracker indicate that about 43% of vegetable oils traded in global markets are affected by some form of export restriction. After implementing a progressive import levy on palm oil in March, Indonesia announced an export ban on crude and refined palm oil, starting April 28, 2022. How long that ban will remain in place is unclear, but restrictions will further compound the import needs of other countries, such as India and Bangladesh, which depend heavily on palm oil to meet their consumption needs.

Export restrictions have also disrupted the soybean oil market. On March 13, 2022, Argentina restricted exports of soybean oil and meal, then altered course on March 31, permitting exports but raising export taxes on those products from 31% to 33%.

All of these factors have contributed to keeping vegetable oil prices at record levels.

Conclusions

As with other key commodities affected by the war, higher prices will have an adverse impact on consumers, particularly those poorest households already threatened by food insecurity.

How long high prices persist will depend much on how quickly peace returns to Ukraine and how long its agricultural economy takes to recover. Under the best of circumstances, however, prices are likely to remain high throughout this year and into 2023. A survey of US planting intentions suggests that US farmers intend to plant a record level of soybean area this year, and early indications suggest that South American soybean production and EU rapeseed production could rebound as well.

The 31 million metric tons of vegetable oil currently converted into biodiesel globally represent an equivalent amount of calories to feed more than 320 million people per year. Thus, suspending biofuel mandates, or restricting mandates to non-edible oils or used cooking oils, could bring welcome supplies of edible vegetable oils to the market, though that is unlikely to happen given the strong political support these policies enjoy. Similarly, export restrictions threaten to exacerbate supply shortages, particularly for key vegetable oils like palm oil. As we saw in 2007/08 and 2010/11, such restrictions can contribute to significant increases in prices. Hopefully, restrictions such as the Indonesian ban on palm oil exports will only be temporary, and will not spread to other countries. If not, higher prices will only make it more costly for the world's poor to meet food security and nutrition needs.

Update

As for other exports, the war has had a serious impact on Ukraine's sunflower-seed oil exports, particularly prior to the re-opening of the Black Sea in August 2022. Through October, Ukraine's sunflower-seed oil exports in 2022 were down about 10% from 2021 levels over the same period. Ukraine's exports of sunflower seeds have been up, however, mostly being shipped by rail for processing in neighboring countries in the West (Romania, Moldova, Germany). Sunflower-seed oil exports account for a significant but relatively small share of the global vegetable oil market (about 13%) compared with other oils such as palm oil, soybean oil, and rapeseed (canola) oil.

Globally, vegetable oil prices have declined significantly from their highs in May 2022, retuning to pre-war levels, in part due to improved rapeseed and soybean production in the northern hemisphere and prospects for a rebound in soybean production in Brazil in 2023, as well as the decision by the Indonesian government to resume palm oil exports in late May. Demand for vegetable oil as a feedstock for biofuel production remains strong. Indonesia has announced plans to increase its biodiesel mandate to 35%, and the use of soybean oil in biodiesel production in the United States continues to grow as well. As a result, vegetable oil supplies will likely remain tight through 2023.

7 High fertilizer prices contribute to rising global food security concerns

Charlotte Hebebrand and David Laborde

Originally published April 25, 2022.

Like people, plants need a multitude of nutrients to thrive. These are categorized into micronutrients, such as zinc and iron; secondary macronutrients, such as calcium and magnesium; and three primary macronutrients: nitrogen (N), phosphorus (P), and potassium (K). Mineral fertilizers provide higher and more plant-accessible nutrients, while organic minerals importantly also provide carbon, which contributes to healthy soils. While efforts to reduce nutrient losses to the environment must be continued and stepped up, it bears emphasizing that fertilizers play a crucial role in agricultural productivity.

World market prices for both food and fertilizer (here we focus only on N, P, and K) increased significantly over the past year and a half and have climbed to even higher levels following Russia's invasion of Ukraine in February, hitting their highest levels yet in March (see Figure 1, +125% from January 2021 to January 2022, +17% from January 2022 to March 2022).

Figure 1: Real prices of food and fertilizer, index

Source: World Bank, U.S.Bureau of Labor Statistics

While there is an immediate concern about the impact of high food prices on food security, especially in low- and middle-income countries, fertilizer price spikes and concerns about availability cast a shadow on future harvests, and thus risk keeping food prices high for a longer period.

In this blog post – which draws on data from the new IFPRI fertilizer dashboard; IFASTAT, compiled by the International Fertilizer Association (IFA); and FAOSTAT – we discuss the underlying drivers of current high prices, explain why the global fertilizer market is particularly susceptible to shocks, and examine which countries are most vulnerable to fertilizer market disruptions. A subsequent post will explore short- and medium-term solutions to address the serious affordability and availability concerns.

Rising prices since June 2020

A combination of factors contributed to fertilizer price increases beginning already at the end of 2020. On the demand side, global mineral fertilizer consumption remained remarkably steady during the first year of the pandemic (+1.6% in 2019/20 compared to the previous crop year); then demand surged by 6.3% in 2020/21. The rebound of crop prices in 2021 contributed to this movement, since higher crop prices tend to increase fertilizer demand as long as farmers face favorable fertilizer/crop price ratios.

On the supply side, the most important factor has been the sharp increase in the price of natural gas, used as both feedstock and energy source in the production of ammonia (a base material for N fertilizers) and which accounts for 70–80% of ammonia production costs. Natural gas prices spiked, especially in Europe and Asia. The large global fertilizer producer Yara announced in March that it was decreasing its European production capacity for ammonia and urea by 55%.

Similarly, rising prices of coal in China, the main feedstock for ammonia production there, along with a phase-out of inefficient production capacity, have also led to reduced production, contributing to rising global fertilizer prices as well. High prices of both ammonia and sulfur, two important inputs for P fertilizers, have driven up prices of P fertilizers. High energy prices, strong demand, and limited supply also drove up spot prices of potash in 2021 and into 2022.

Other supply side drivers, such as ongoing COVID-19 related bottlenecks in global value chains and the impacts of hurricanes on production centers in the US Southeast, have also disrupted production and transportation of both fertilizers and fertilizer inputs.

The Ukraine conflict, the economic sanctions it triggered, and disruptions in the Black Sea trading routes have all further increased trade costs and uncertainty about Russian and Belarusian exports – a particular problem for the international fertilizer trade. In 2020 Russia was providing 14% of globally traded urea (the most widely applied N fertilizer), 11% of MAP and DAP (the most widely applied P fertilizers), and jointly Russia and Belarus contributed a whopping 41% of all traded MOP (the most widely applied K fertilizer). Finally, as Europe faces continued shortfalls in natural gas sources, production costs will remain high for all fertilizer suppliers (except Russia) including for the United States and for Persian Gulf countries manufacturing N products.

Why the global fertilizer market is particularly vulnerable to supply shocks

All of these problems have been amplified by two factors: A large amount of fertilizer – 38% of all N, 50% of all P, and 80% of all K produced – is traded on international markets, and the lion's share of that trade derives from only a few countries (Figure 2). Since ammonia, the key ingredient in N fertilizers, is primarily produced using gas or coal as feedstocks, many countries could in theory manufacture them, but comparative advantage lies with countries with relatively lower gas or coal prices. In 2019, Russia, China, and Qatar together accounted

Figure 2: Global fertilizer markets, 2019, million metric tons

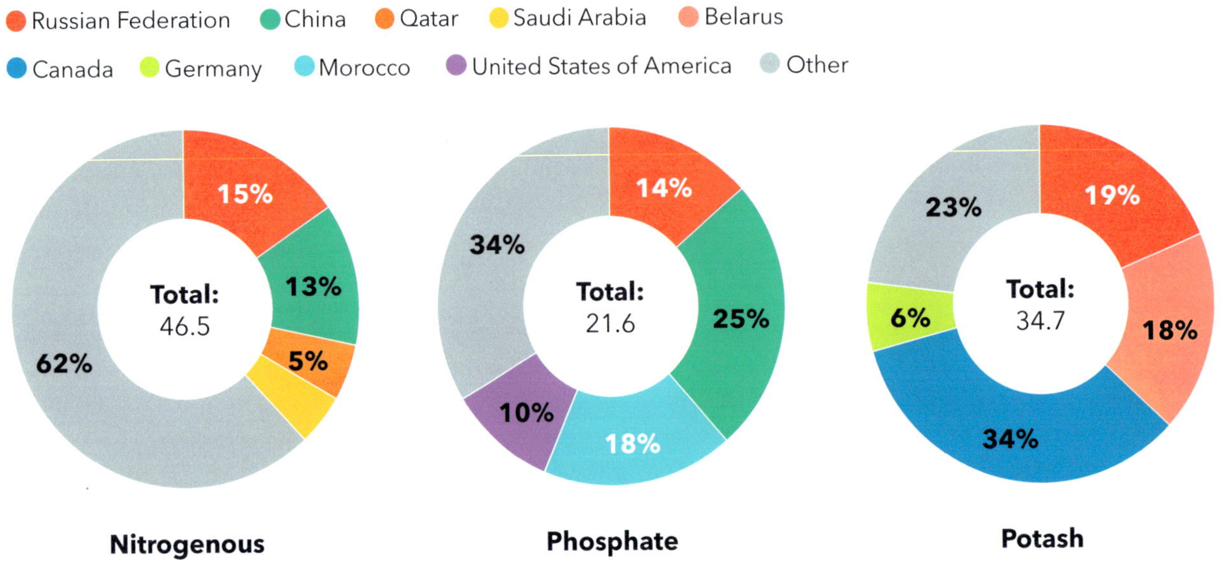

Note: Global production of N is estimated at 123 mio MT, of P2O5 44 mio MT, and of K2O 44 mio MT.
Source: FAOSTAT

for 33% (15%, 13% ,and 5% respectively) of N traded. The production of potash and phosphates is more concentrated due to the uneven distribution of source deposits: For phosphates, the top three exporters, China, Morocco, and Russia, together represent 57% of global trade (25%, 18%, and 14% respectively), and for potash the market share of the top three reaches 80% (Canada 39%, Russia 21%, and Belarus 20%). The role of Russia in N, P, and K markets, in conjunction of Belarus for potash, is particularly important in the current crisis.

In this context, export policies of key countries are critical in shaping market stability, and all eyes are on China to see whether, in April 2022, it will lift export restrictions on N and P imposed in October 2021. The export restrictions are not limited to China; IFPRI's Food and Fertilizer Export Restrictions Tracker shows five countries have imposed such measures, collectively reducing the global trade of each individual nutrient by 20% (see the IFPRI fertilizer dashboard).

These export restrictions are having serious impacts on a number of vulnerable countries: Mongolia, trapped between Russian and Chinese policies, has seen 98% of its fertilizer supply restricted, Nicaragua 80%, Ecuador 74%, Côte d'Ivoire 63%, Cameroon 60%, and Azerbaijan 43%, among others.

Another issue is trade remedies, that is, anti-dumping duties applied by importing countries that can impact prices in those markets. Examples include remedies applied by the European Union and the United States, already implemented on phosphates, and under consideration for urea ammonia nitrate (UAN).

Repercussions of record fertilizer prices

Clearly, countries depending heavily on fertilizer imports from Russia and Belarus (Figure 3, or the Fertilizer Dashboard for details by nutrient) face an immediate shortfall, and will need to secure alternative sources

Figure 3: Share of fertilizer imports in domestic use

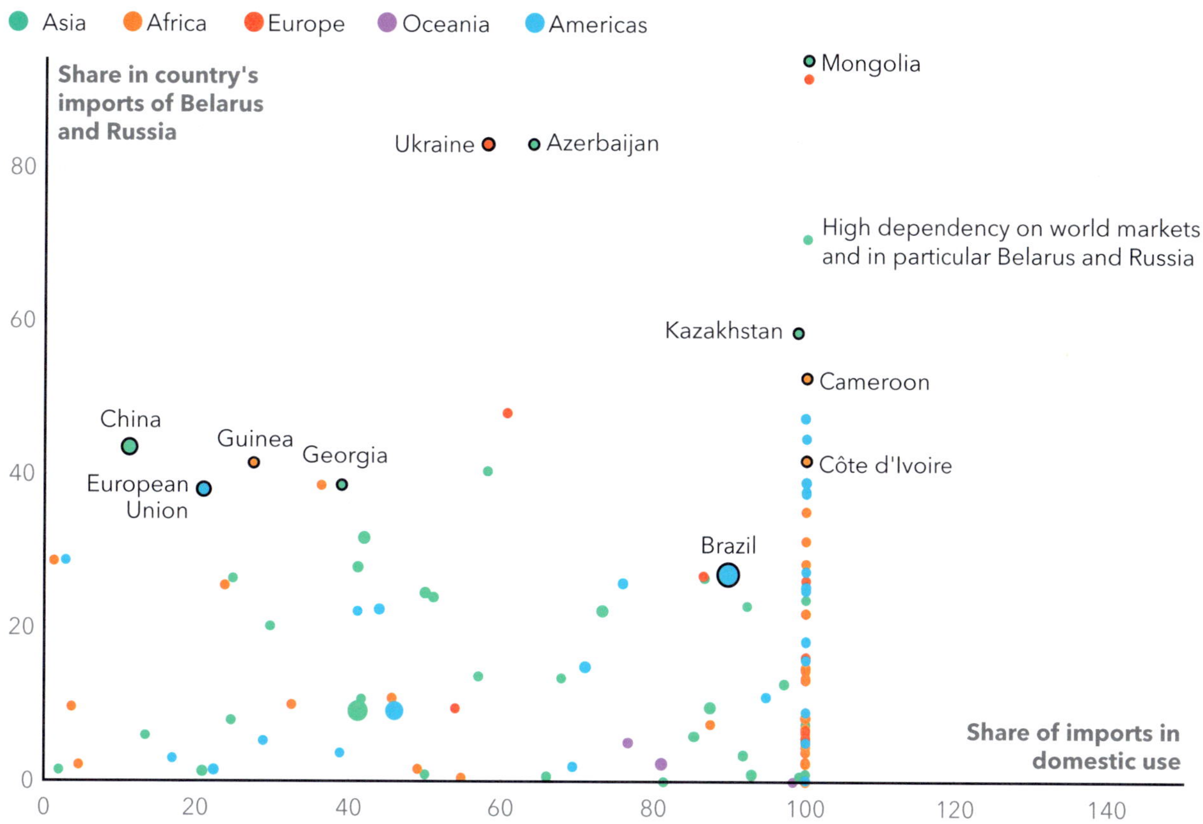

Note: Size of bubbles represent the imports of a country in metric tons of nutrients

Source: FAOSTAT

from a very tight global market. If they are able to procure supplies from other exporting countries, greater distances and transportation rerouting will lead to additional costs, pushing fertilizer prices even higher and/or lead to lack of availability. With three-quarters of countries depending on imports for 50% or more of their fertilizer use, perturbations on world markets will spread fast, impacting countries that do not import directly from Russia or Belarus.

These problems will spread in the coming months; ultimately all farmers around the world will be impacted. The price shock may be buffered for farmers in some emerging and developing countries that have fertilizer subsidy regimes in place (such as India, China, Bangladesh, and Ghana), but those regimes are going to place tremendous fiscal pressure on budgets already stressed by substantial government outlays during the COVID-19 epidemic.

Relatively smaller markets – especially many African countries – face a particularly difficult situation, as producers and traders are likely to favor shipping limited supplies to larger markets. Given Africa's still-limited use of fertilizers – an estimated average of 25 kg/ha, a fraction of the global average of 121 kg/ha – a decline in fertilizer use would lead to significantly reduced productivity for the continent, with potentially serious consequences for food security.

Finally, many fertilizer manufacturing plants in Africa focus on blending imported components to deliver the required N, P, and K compounds required by local farmers. Canceled or delayed deliveries of only a single input, for instance potash, could disrupt an entire operation, depriving farmers of inputs for weeks or months, until alternatives are put in place (for example, application of a single nutrient).

Conclusions

While risks around fertilizer availability and affordability will vary by country and region, farmers in many low-income countries face potential hardship as the Ukraine crisis continues. If these challenges are not addressed, harvests will suffer, keeping pressure on food prices beyond the short run and adding to food security concerns in many developing countries. A subsequent post will examine a suite of potential steps related to plant nutrition to address the crisis in the short term and prepare for a more resilient and sustainable future.

8 The Russia-Ukraine war after a year: Impacts on fertilizer production, prices, and trade flows

Charlotte Hebebrand and Joseph Glauber

Originally posted March 9, 2023.

Russia's 2022 invasion of Ukraine triggered global disruptions in markets for key food crops and fertilizers, threatening food security worldwide. With the war now entering its second year, high international food prices have moderated, though domestic price levels remain high in many low- and middle-income countries. Here, we turn to global fertilizer markets, examining how they weathered the past year, and the likely impacts going forward for agricultural production and food security.

The good news is that fertilizer prices, while still high, have fallen significantly from their 2022 peaks (Figure 1). As with the global trade in food crops, that rise and fall is a complex story of shifting supplies and trade linkages.

Figure 1: Fertilizer prices, US dollars per metric ton

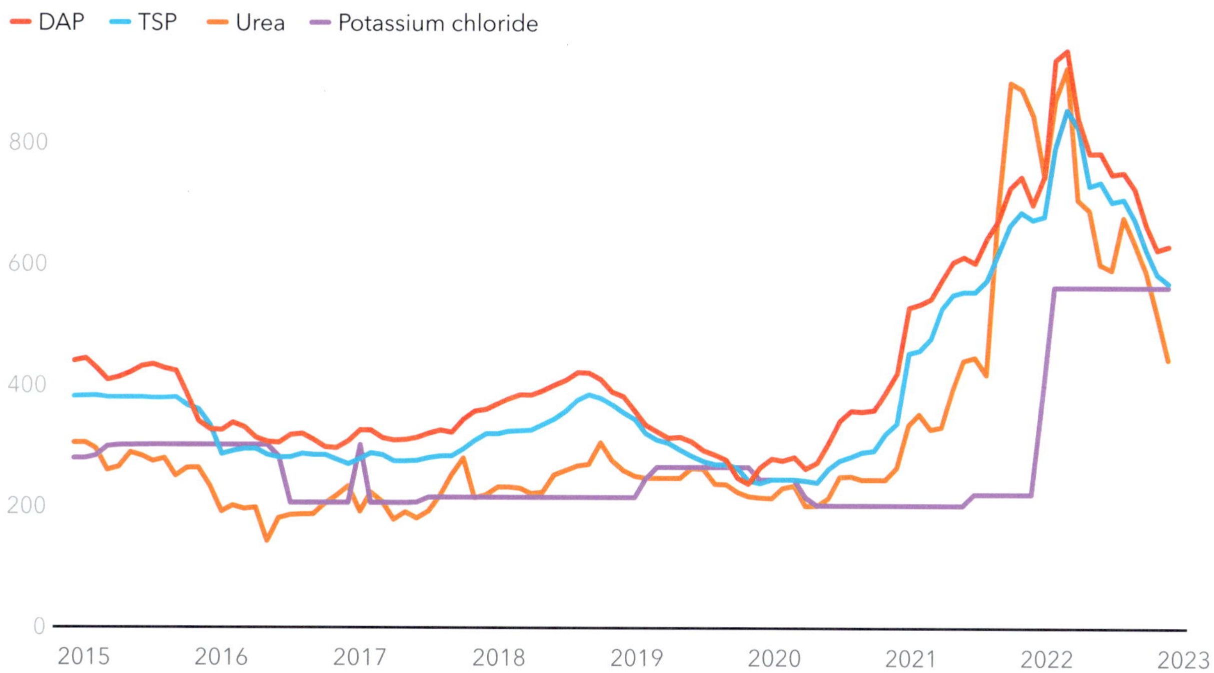

Note: DAP = diammonium phosphate; TSP = triple super-phosphate
Source: World Bank

At the time of the Russian invasion, fertilizer prices were already at historically high levels. They had been rising since late 2020 due to several factors. Fertilizer demand, which declined during COVID-19 lockdowns, rebounded in late 2020/21 as restrictions were lifted and crop prices rose. On the supply side, increases in prices of natural gas and coal – key feedstocks and energy sources in fertilizer production – as well as some reductions in production capacity also added upward pressure on prices.

Then the outbreak of war on February 24, 2022, drove prices dramatically higher. Uncertainties multiplied about fertilizer exports from Russia and Belarus due to the conflict itself, new or expanded economic sanctions on the two countries, and disruptions in Black Sea trade routes. Russia and Belarus are important producers of all three major fertilizer nutrients: In 2020, Russia accounted for 14% of global trade in urea and 11% of trade in phosphate, while jointly Russia and Belarus accounted for 41% of global trade in potash. The fact that a small number of countries produce a large share of internationally traded fertilizers makes the sector vulnerable to trade shocks.

Countries heavily dependent on fertilizer imports from Russia and Belarus feared an immediate shortfall, and many had to scramble to secure alternative sources from a very tight global market. But with some three-quarters of all countries importing at least 50% of their fertilizer consumption (Figure 2), trade shocks created by the

Figure 2: **Share of fertilizer imports in domestic use**

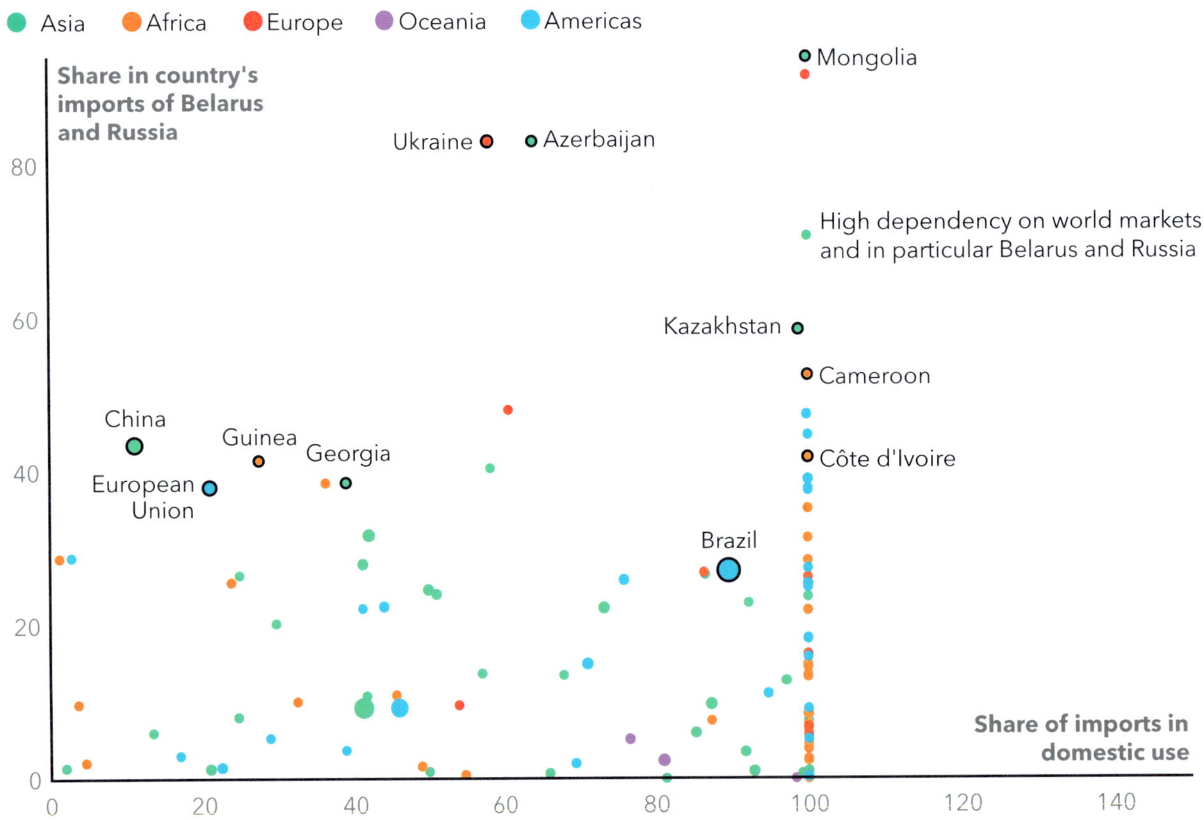

Note: Size of bubbles represent the imports of a country in metric tons of nutrients

Source: FAOSTAT

war reverberated around the world. Post-invasion turmoil in natural gas markets and skyrocketing gas prices also contributed to rising fertilizer prices.

Sanctions, export restrictions, and changes in trade flows

The sanctions imposed by the European Union, United States, Canada, and other countries on Russia and Belarus after the invasion – on top of earlier restrictions – formally exempt agricultural products. EU sanctions do ban potash imports from Belarus and forbid shipments of Belarusian potash through EU territory to other markets. Despite the carve-outs for agricultural products, sanctions may have led to decreased fertilizer trade to some regions, since importers may choose not to purchase from these two countries because of added costs of doing business with them, such as more restrictive banking regulations, higher insurance costs, or out of fear of being ensnared in financial sanctions.

An IFPRI analysis estimated potash exports from Belarus were at least 50% lower in 2022 than in 2021 because of the sanctions and restrictions on using EU territory for transit. While total Russian urea and potash exports were down in January-August 2022 compared to the same period in 2021, they recovered in the remaining part of the year. The shutdown of the Tolyatti ammonia pipeline to Odesa contributed to a dramatic decline of Russian ammonia exports – 63% for the period January-August 2022 over 2021 (some of the decline might also be linked to "overcompliance" with sanctions).

Also disrupting global markets were various types of export restrictions imposed by some countries keen to keep domestically produced fertilizer available for domestic consumption. These included outright bans or onerous inspection and licensing processes. IFPRI estimated in June 2022 that some 20% of global fertilizer trade was impacted by such restrictions.

In particular, fertilizer exports from China plunged after the country restricted exports from mid-2021 onward. Chinese exports of diammonium phosphate (DAP), which typically account for 30% of global DAP trade, fell by 43% in 2022 compared with 2021, while Chinese urea exports declined by 47% in the same timeframe (Figure 3).

Large fertilizer importers facing a shortfall from Russia and Belarus were able to secure supplies from alternative sources. Brazil, for example, the second largest importer of potash, managed to increase imports from Canada to help offset the decline from Belarus; Morocco, the fourth largest global ammonia importer, stepped up imports from Saudi Arabia and Egypt to make up for shortfalls from Russia. Meanwhile, production capacities of some regions increased along with exports – in particular potash from Canada and phosphate from Morocco, as well as urea from Nigeria. At the same time, Russian fertilizer exports also increased dramatically in some markets, such as India.

Figure 3: China's fertilizer exports, metric tons

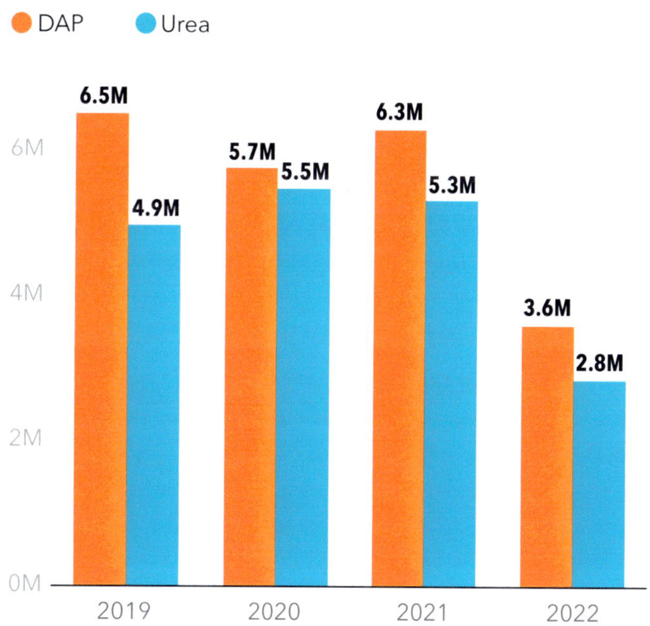

Note: DAP = Diammonium phosphate

Source: General Administration of Customs, People's Republic of China

However, some smaller low-income countries in sub-Saharan Africa (SSA) found it much more difficult to secure fertilizers, as they encountered problems with fertilizer availability after the invasion.

While the International Fertilizer Association (IFA) estimates a global decline in fertilizer consumption of 5% for fertilizer year 2022, the International Fertilizer Development Center estimates that fertilizer consumption in SSA, outside of South Africa, may have declined by as much as 25%, putting the continent back to consumption levels seen in the mid-2010s.

Over the course of 2022 and into 2023, international prices eased and fertilizers became more available in global markets, but remained unaffordable in many African countries due to persistently high domestic price inflation. Even without the price pressure, fertilizer prices in Africa are generally higher than in the rest of the world, given the still significant transportation infrastructure and regulatory bottlenecks.

Fiscal constraints and a lack of foreign exchange created difficulties for financing more expensive fertilizer imports. Subsidy bills ballooned in some countries, and in some cases became unsustainable. However, assistance from global organizations has made a difference. International development banks including the World Bank and the African Development Bank offered financing facilities; and other initiatives, such as Sustain Africa, also supported by donations from the fertilizer industry, sprang up to help channel fertilizers to African countries in need. The World Food Programme, which traditionally provides food aid, also offered its services on a full-cost recovery basis to ship fertilizers donated by a Russian fertilizer company.

Possible impacts on agricultural production and food security

Assessing the impact of the fertilizer crisis – in particular, reduced demand – on crop yields is a complex undertaking, especially in regions of the world with large numbers of smallholder farmers. Yield impacts vary depending not only on amounts being used but on shifting fertilizer choices. In times of high prices, farmers tend to favor nitrogen, and thus there has been a steeper global decline in demand for potash and phosphate. Yield impacts from decreased nitrogen use can be witnessed within the same growing season, while the impacts of skimping on potash and phosphate on yields, and also on soil health, may take several years to materialize.

Nor is it easy to gauge impacts of reduced fertilizer consumption on food security. Many factors affect production: climate- and weather-driven events, such as droughts and floods, or high postharvest food losses can drive down productivity even if sufficient fertilizers have been applied. Fertilizers are also used to produce nonfood crops, particularly for biofuels. Global biofuel production is still increasing, albeit at lower rates since 2020, and by and large is still relying on food crop feedstocks rather than advanced nonfood crop feedstocks (Figure 4). For example, almost 38% of the US corn crop is used as feedstock for biofuel production, and globally, vegetable oils are forecast to account for some 23% of biofuel feedstocks by 2027.

The allocation of fertilizers in Africa, tracked for some key countries by AfricaFertilizerWatch, also merits close observation. High fertilizer prices and limited supplies may push farmers to further prioritize cash crops over food crops. Cash crops such as coffee, tea, sugar, and tobacco account for a significant share of overall domestic use of fertilizers, going as high as 80% in Côte d'Ivoire and Uganda, and 90% in Mozambique. Cash crop farming can boost economic livelihoods, but does not boost food crop production. Thus it will be important to assess how farmers distributed their fertilizers in 2022.

Lower international prices, but what comes next?

The high prices of 2022 led to reduced demand (or demand destruction) at the farm level, which in turn has brought prices down from their peaks of last year. But following two consecutive years of an overall 7% decline, IFA forecasts a global increase in fertilizer demand of 3% in fertilizer year 2023.

Figure 4: **Biofuel demand share of global crop production**

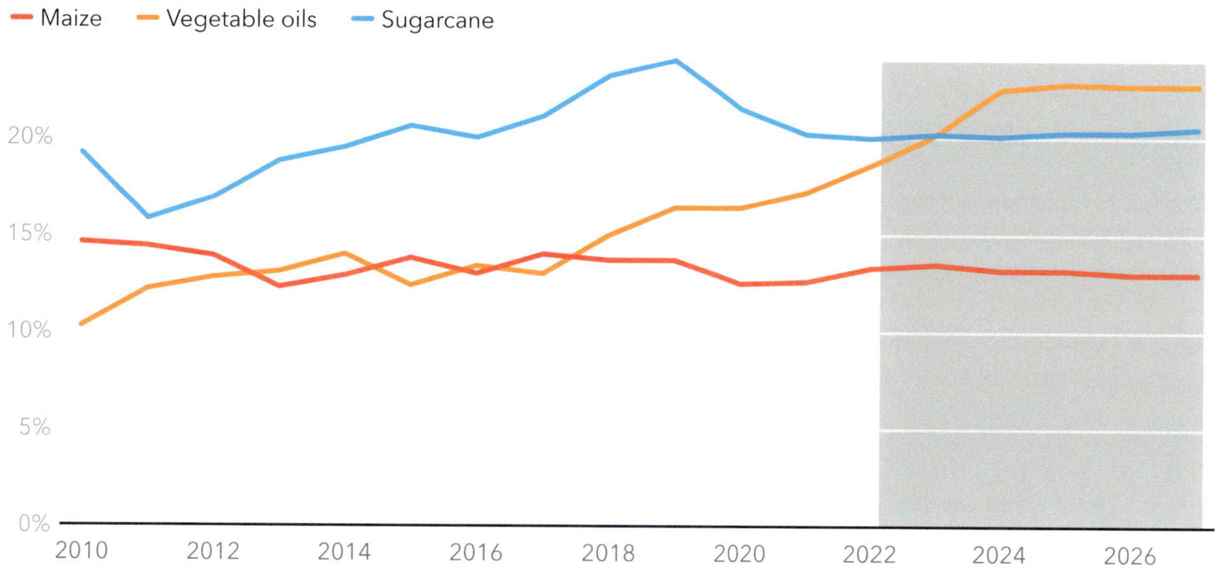

Note: Projections (shaded area) reflect main case scenario.
Source: International Energy Agency

Prices remain high by historical standards, however, and domestic price inflation also persists in many countries. The fertilizer sector thus remains vulnerable to further shocks arising out of the Russia-Ukraine war or other developments in an uncertain global market environment.

Capacity increases can ease supply concerns, but greenfield projects require large capital investments and long timespans to come to fruition. In 2023, new ammonia capacity is expected to come online in 2023 in the United States. Easing China's export restrictions would also have a significant impact on global markets. Re-opening the Togliatti pipeline would bring more Russian ammonia to global markets, but such a move depends on Ukrainian approval. Ongoing negotiations on the renewal of the Black Sea Grain Initiative (which expires on March 18) under UN auspices also include discussions on the pipeline. Exports of Belarusian potash could conceivably also increase in 2023 if some EU sanction carve-outs were put into place or efforts made to end over-compliance.

While the outlook for fertilizer year 2023 is brighter than that of 2022, the fertilizer sector by its nature remains vulnerable to trade and energy shocks, and the war and other global problems continue to pose serious risks.

Meanwhile, the sector also faces significant longer-term issues that can get lost amid the contingencies of crisis response. These include the huge reliance on fossil fuels for ammonia production and the search for viable alternatives (such as green ammonia produced from electrolysis powered by renewable energy). Fertilizer continues to be used inefficiently in many regions of the world, with negative impacts on soils, water, and climate change. This problem can be addressed through improved fertilizer application, specialty and biological fertilizers, and other solutions. Organic fertilizers require more innovation and promotion for widespread use, as does integrated nutrient management that combines organic and mineral fertilizers. Plant nutrition should also be better integrated with soil health, and it is encouraging that the African Union has opted to hold a Fertilizer and Soil Health Summit in Senegal in June 2023. Overall, countries and international organizations are well advised to find a balance between crisis-oriented fixes and longer-term industry transformation.

9 Is food price inflation really subsiding?

Rob Vos, Joseph Glauber, and David Laborde

Originally published January 24, 2023.

Almost a year after the February 2022 Russian invasion of Ukraine, fears of a period of sustained high global food prices have subsided somewhat, but eight major concerns for food security remain.

First, prices for food commodities remain high by historical standards. As shown in Figure 1, the UN's Food and Agriculture Organization (FAO) index for internationally traded food commodities, after rising during the initial months of the war, is now back to its pre-war, end of 2021 level – but still remains well above levels of preceding years. This holds for all main components of the index (cereals, meat, dairy, vegetable oils, and sugar). Vegetable oil prices showed the strongest decline in the second half of 2022, falling by 33% between June and December, but still are one-third above pre-COVID levels.

Figure 1: Nominal FAO Food Price Index, 2019-2022

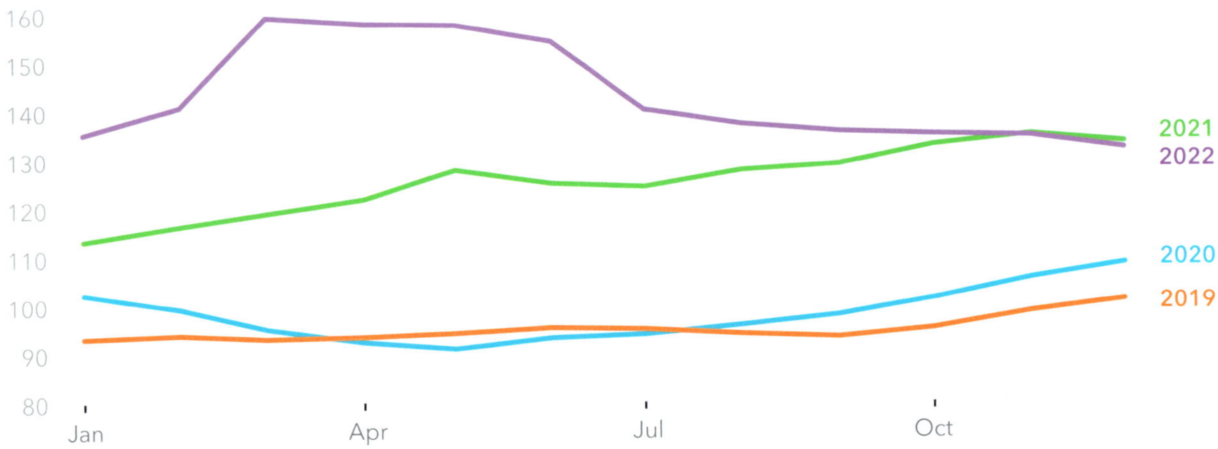

Source: FAO

Second, markets for staple foods remain tight. Global stock-to-use ratios for grains remain at around or below the lows of recent years (Figure 2). This indicates price instability is likely to intensify with any major supply shock, especially when also considering the ongoing uncertainty around the actual availability and exportability of grains stored in Ukraine as the war continues.

Third, after disrupting exports of Ukraine's 2021 and 2022 crops, the war now threatens Ukraine's 2023 commodity supplies. Fall plantings of wheat were down by as much as 40% from 2022 levels, and the conflict will likely have a significant negative impact on spring plantings.

Figure 2: Global ending stocks (excluding China), days of use

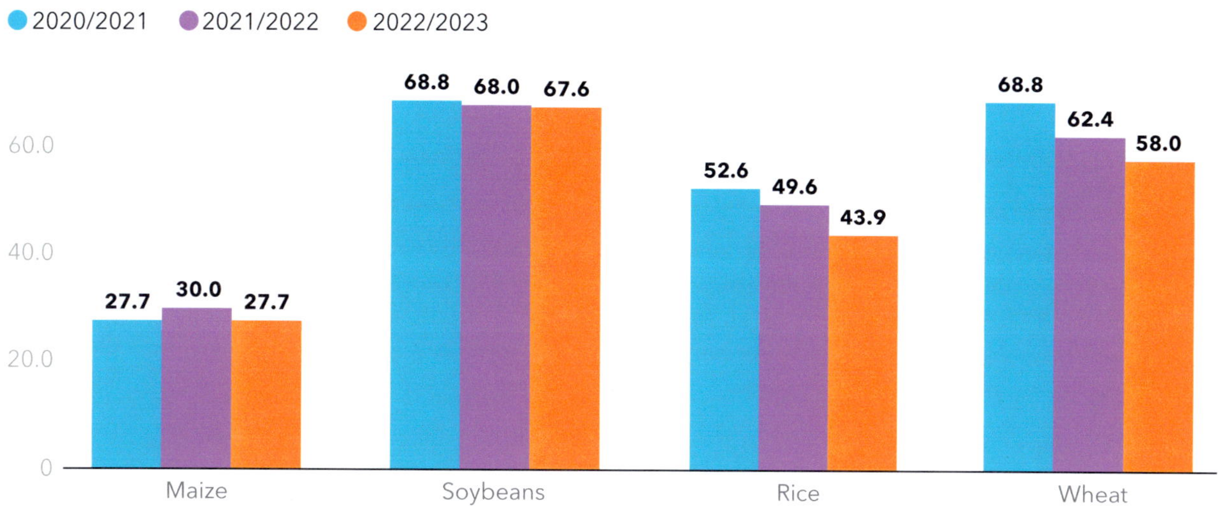

Source: USDA/FAS PSD online (February 8, 2023)

Fourth, fertilizer prices have fallen from their peaks but also remain high, despite recent drops in natural gas prices (a major input for nitrogenous fertilizer as well as an energy source for the production process), as shown in Figure 3. Constraints on supplies of potash, including the ban of exports from Belarus, are keeping upward pressure on prices. With output prices for staple foods falling, these high input costs are depressing farm profitability. This likely will depress fertilizer usage, which in turn will affect yields, especially of rice, wheat, and maize. A first concern is with rice; production dropped somewhat in 2022 and prospects of lower stocks have led to a surge in prices for rice in futures markets.

Figure 3: International fertilizer price index by main nutrient contents

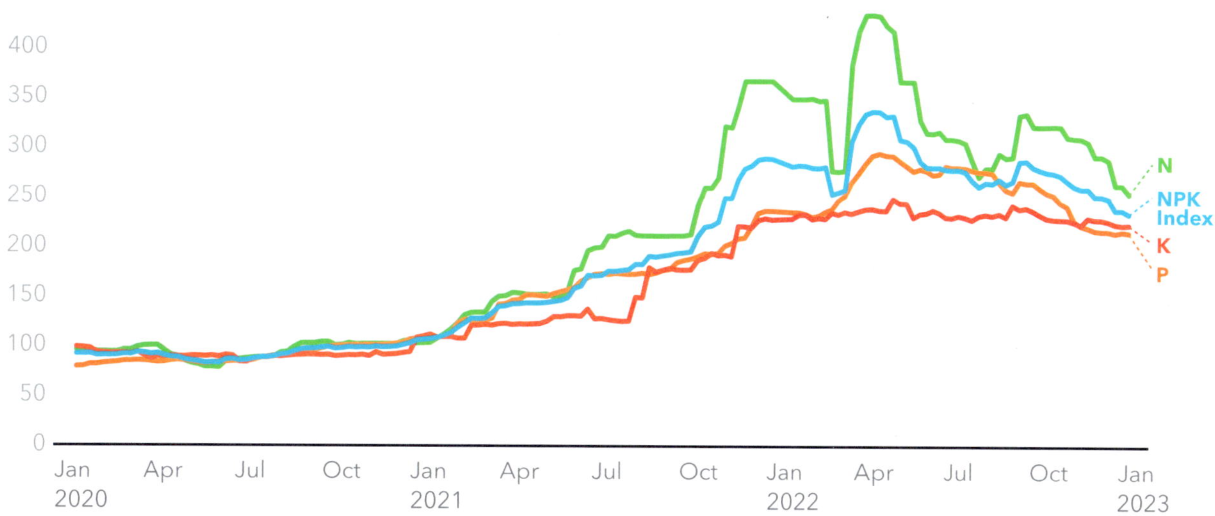

Source: IFPRI based on Bloomberg product level data

Fifth, adverse climatic conditions in the southern hemisphere have overlapped with prolonged drought conditions in Argentina and East Africa, resulting in sharply reduced 2023 production prospects for wheat and other crops compared to last year. Abnormally wet but favorable conditions in Australia and strong maize production in Brazil could compensate for some of this reduced cereal production (AMIS Market Monitor, December 2022).

These overall trends, together with high fertilizer prices, mean reduced prospects for global supplies of staple foods in 2023, even if – with no end to the war in sight – the recently extended Black Sea Grain Initiative remains in effect during the year.

Sixth, global demand for food may well weaken further during 2023, given the expected significant slowdown of the global economy. Without any more major disruptions, further global food shortages are not expected to emerge in 2023, though there could be shortages in local hotspots – such as in Afghanistan and the Horn of Africa – where conflict, new weather shocks, and lack of import capacity may continue to affect food availability (see WFP-FAO hunger hotspot early warnings).

Seventh, the drop in world market prices of staple foods during the second half of 2022 has at best slowed domestic food price inflation very modestly in most countries. Consumer prices for both food and other essentials have continued to increase sharply in most parts of the world (Figure 4). As analyzed by Martin and Minot (2022), shifts in international prices have limited effects on domestic prices. For instance, on average, less than half of the rise in global wheat prices was passed on to higher domestic wheat prices during the surges

Figure 4: Consumer price inflation rates, November to December 2022

(percentage change year-over-year)

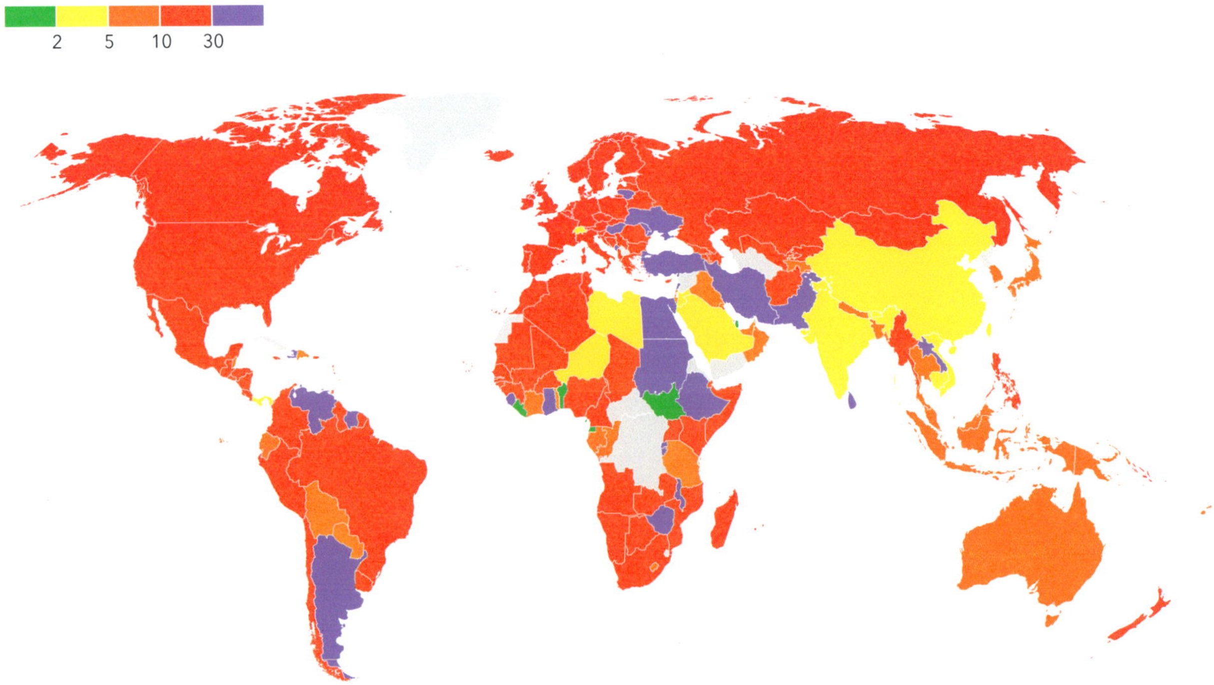

Source: Food Security Portal based on data from IMF and Trading economics

in 2021 and 2022. The pass-through is even lower when prices fall. When international wheat prices dropped by about 28% between May and July 2022, domestic prices for wheat products declined by only 8% on average. The low degree of transmission of global market prices of course relates to food import dependence, but to a large extent also to insulation policies, which can include export restrictions, lower import tariffs, and domestic food subsidies. While these policies have protected consumers from the impacts of the global food price shocks, they also contributed to the food price surges in the first half of 2022 (see Martin & Minot 2022).

Eighth, and most critically, low-income countries face ongoing macroeconomic problems that pose additional risks to food security. The capacity of governments to protect consumers from both international and domestic food price shocks is weakest in low-income countries, where the need for protection is greatest. Many people in such countries spend significant portions of their incomes on food, and as prices rise rely on forms of government support – yet after repeated shocks, public budgets are already over-extended. Efforts to mitigate the impacts of the COVID-19 pandemic on livelihoods have driven up government debt to unsustainable levels. The share of low-income countries in debt distress has doubled to 60% since 2015 (IMF 2022). Thus, high debt service obligations along with rising food and energy import prices have both constrained import capacity and weakened the currencies of these countries, which in turn has fanned domestic food price inflation.

Consequently, food affordability remains a challenge at both the macroeconomic level (capacity of countries to pay their food import bills) and at the household level (increased prices in local currency for imported food and energy, and reduced prospects to raise incomes given the prospect of a global economic slowdown). These risks will remain high going forward; without major debt relief and additional financial support from the international community, food prices are likely to continue to climb in 2023, posing a continuing threat to the food security of vulnerable households around the world.

10 Assessing tight global wheat stocks and their role in price volatility

Joseph Glauber

Originally posted March 14, 2023.

The Russia-Ukraine war has caused significant price volatility in agricultural markets over the past year – for wheat, in particular, price levels and price volatility reached the highest levels since the 2007/08 marketing year. Both have fallen back to pre-war levels over the past six months, but volatility remains high relative to historical levels (Figure 1), indicating that significant market uncertainty remains, creating ongoing vulnerability for global food security.

Before 2022, volatility (measured by the implied volatility in the wheat futures market) spiked and remained high three times since 2006 – in 2007/08 and then again in 2010/11 and 2012/13 (as indicated in the gray bands in the figure). Markets then remained relatively quiet until the recent spike following Russia's invasion of Ukraine in February 2022. High volatility persists due to market uncertainty over the war and the relative tightness in stock levels, which provide little cushion against unforeseen production shortfalls.

Figure 1: Wheat price volatility

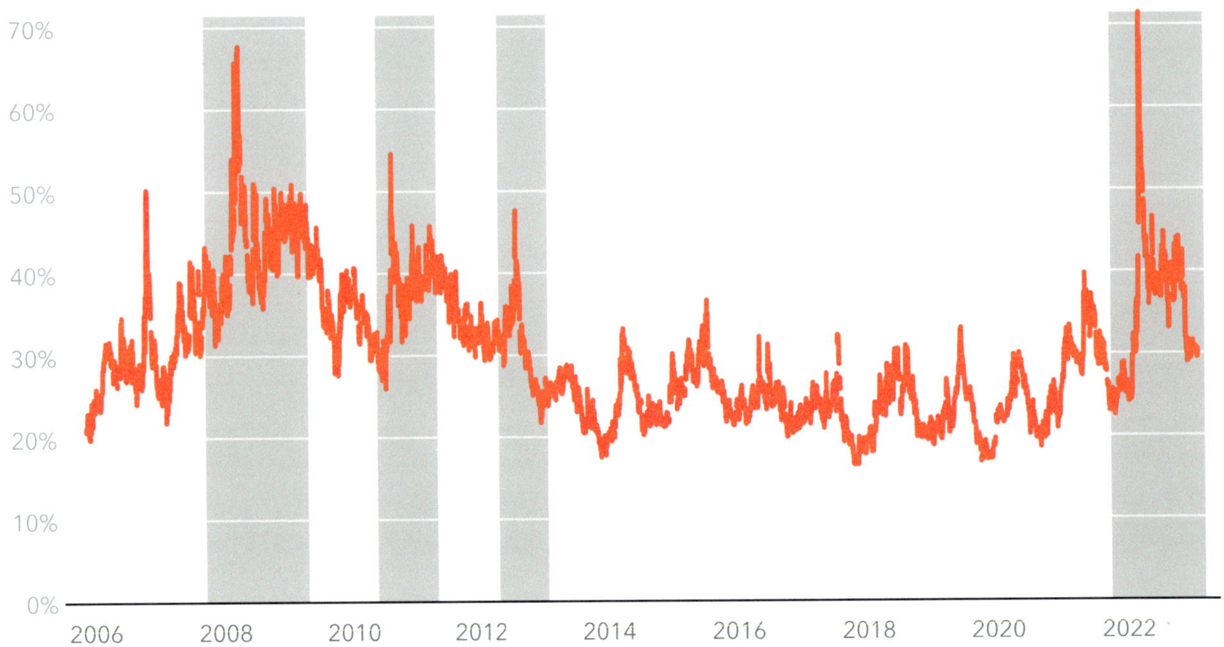

Note: Volatility is measured as the implied price volatility in the underlying futures/options market.

Source: Bloomberg

Price volatility can, in part, be traced to uncertainty over the continuing flow of supplies, which depends on a number of factors, principally current production and existing stocks. This blog post focuses on the latter, examining the link between global wheat stocks and price volatility, as well as how – and how much – the tightness in global stocks over the past 18 months has exacerbated volatility, and prospects going forward.

Explaining global stocks

Stocks are crucial for buffering the impacts on market prices caused by production shocks such as droughts, extreme weather, conflicts, or other disruptions. When production shortfalls occur, available inventories can help to tide markets over until the next harvest, thus moderating price impacts. Conversely, when global harvests are large, wheat can be carried forward for sale in the next marketing year, thus putting an effective floor on current prices. In this way, stocks help moderate price peaks and troughs and thus reduce price volatility. However, when stocks are at already low volumes, as they have been over recent months, the capacity to buffer the impacts of shocks is more limited.

Tracking global wheat stocks, and their connection to market price volatility, is complicated. Measurements are typically based on aggregating country ending stocks (often called carryover stocks), the amount held at the end of a marketing year before the new crop is harvested and available for consumption. A country's ending stocks are equal to the previous year's ending stocks, plus any new production and imports, minus what is consumed domestically and exported during the year. In general, world stocks refer to the global sum of ending stocks at the close of each country's national marketing year. Because the dates of marketing years may vary across countries due to differences in local crop calendars, estimates of global ending stocks are not an actual measure of available stocks at a particular point in time, but serve more as an indicator of the level of current supplies that will be carried into the next marketing year.

Countries with public stockholding programs account for about 65% or more of global wheat stocks. Public stocks are held by governments for food security purposes, to bolster producer income, or both. Globally, the balance of publicly held stocks has shifted in recent years from the United States and European Union (EU) to other countries, notably China and to a lesser extent India. Historically, the United States and EU maintained large wheat reserves as part of their price support programs, but policy reforms in the 1980s and 1990s eliminated those programs. More recently, public stockholding schemes like the Food Corporation of India and the Chinese state-owned food company COFCO have bought, sold, and stored large quantities of wheat and other grains.

The remaining one-third or more of global stocks are held by private and commercial entities (including individual farmers). A distinction is often made between private stocks held for discretionary purposes (sometimes called "speculative" stocks) and "pipeline" or "working" stocks. Discretionary stocks are held by commercial interests that purchase grain in one period with the hope of selling later at a price high enough to recoup the costs of purchasing and storing the crop. Pipeline stocks are stocks held for operation of the marketing chain and are generally less price responsive than discretionary stocks.

Unfortunately, the diversity of governments and private entities involved means data on global stocks are incomplete and often of poor quality, meaning it is difficult to have complete confidence in world ending stock figures. Many countries (including some developed economies) fail to collect or publish data on stocks, and thus analysts must impute the level of stocks based on production and consumption data. While improving the quality of stock data has been the focus of organizations like the Agricultural Market Information System (AMIS), much uncertainty remains over data quality. In addition, the physical quality of the stocks is often unknown. If stored properly, wheat can be held for several years, but if the quality deteriorates sufficiently, it may be unusable for milling purposes and must be used as animal feed or for industrial uses like biofuels.

How do stocks affect price volatility?

Typically, stocks serve to limit potential price impacts from rises and falls in production. In the event of a bumper crop, private storers purchase grain with the intention of selling it in the next period for an expected profit. This effectively puts a floor on market prices. Likewise, when stocks carried forward from the previous year are high and there is a production shortfall, storers will sell grain rather than carrying it forward to the following year. This helps moderate prices and buffers the impact of the production shortfall.

But the current global market environment – with its combination of post-pandemic effects, droughts, impacts from the war, and other issues – is anything but typical, and has led to generally tighter stocks. When stocks are tight, the ability of storers to buffer production shocks is limited. When discretionary stocks are limited and a production shock occurs, holders of pipeline stocks (such as millers) are reluctant to release the grain to ensure adequate supplies for milling and other users. Yet consumers often are willing to pay more for scarce grain supplies, giving up other expenditures (see blog post on the impacts of food price shocks on diets of Egyptian households). Thus, prices keep rising; only large price increases can reduce demand sufficiently to meet the full impact of a supply shock. As a result, the price volatility tends to be far higher when stocks are low.

How tight are wheat ending stocks?

Understanding the risks posed by current volatility depends on gauging the size of global stocks relative to global consumption. The U.S. Department of Agriculture estimates that global wheat ending stocks for the 2022/23 marketing year will be 267 million metric tons (Figure 2). More than half of these stocks are estimated to be held by China, with the EU, United States, and other major exporters accounting for another 20%. China's wheat stocks increased by over 160% between 2012 and 2020. This was largely due to changes in China's agricultural policy, which increased producer support prices and resulted in the accumulation of large government stockpiles. By contrast, wheat stocks held by the rest of the world declined by 12% over same period.

Some analysts exclude China's stocks when gauging market tightness. The main rationale for doing so is that data on China's stocks are not reported by the government and thus highly uncertain. Second, China's stocks are largely in government control and isolated from global markets (indeed, despite large stockpiles, China continues to be a large importer of wheat). Lastly, there are many concerns about the quality of the wheat in those reserves. Excluding China, global wheat ending stocks are estimated to be 128 million metric tons for 2022/23.

A common measure of the relative tightness of stocks is the stock-to-use ratio (SUR), which expresses ending stocks as a percent of consumption. The SUR allows one to normalize stock levels by accounting for changes in consumption due to growth in population and income, and other factors, thus allowing a more consistent comparison across years.

To assess the relative tightness of current stocks, we consider three separate stock measures:

- **The first measure is based on total global stocks.**
- **The second measure is based on total global stocks excluding China.**
- **The third measure considers just those stocks held by the major exporters** (the United States, EU, Argentina, Australia, Canada, Kazakhstan, Russia, and Ukraine), and is often used by market analysts to gauge stock tightness.

All of these measures are divided by global wheat consumption, and the resulting SUR is then multiplied by 365 days to express stocks as days of use.

The three measures track very closely from 2000/02 to 2012/13 (Figure 3).

Figure 2: Global wheat stocks, metric tons, 2000/01–2022/23

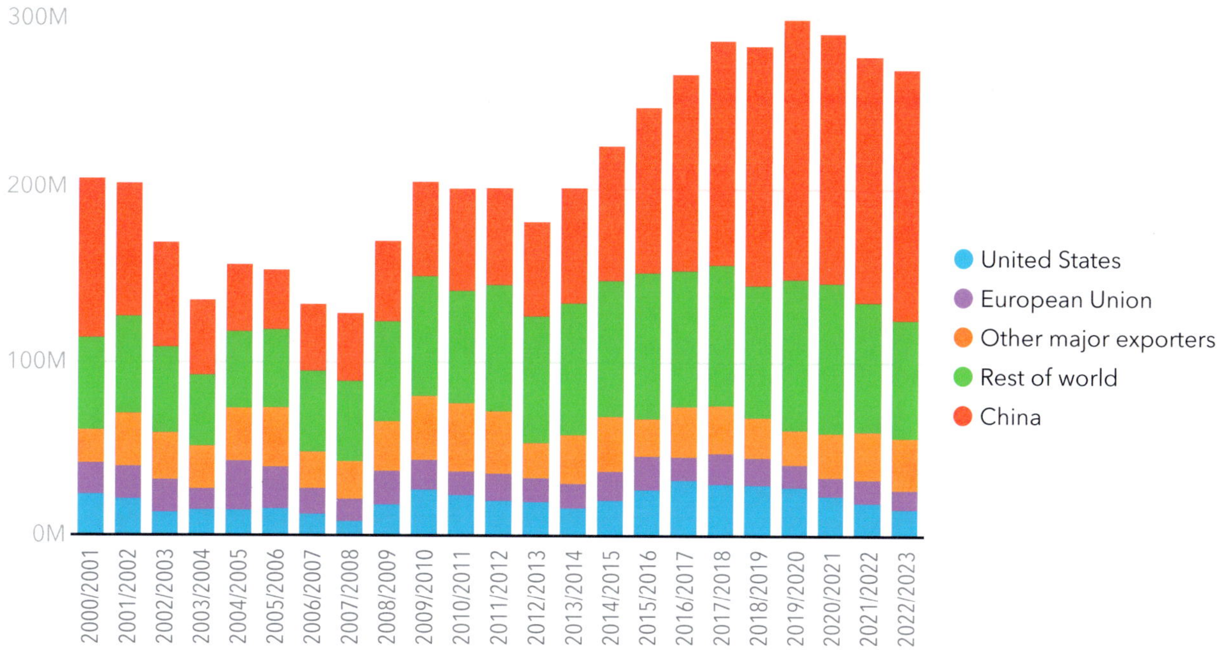

Note: Other major exporters include Argentina, Australia, Canada, Kazakhstan, Russia and Ukraine.

Source: USDA/FAS PSD database

Figure 3: Global wheat stocks, days of use, 2000/01–2022/23

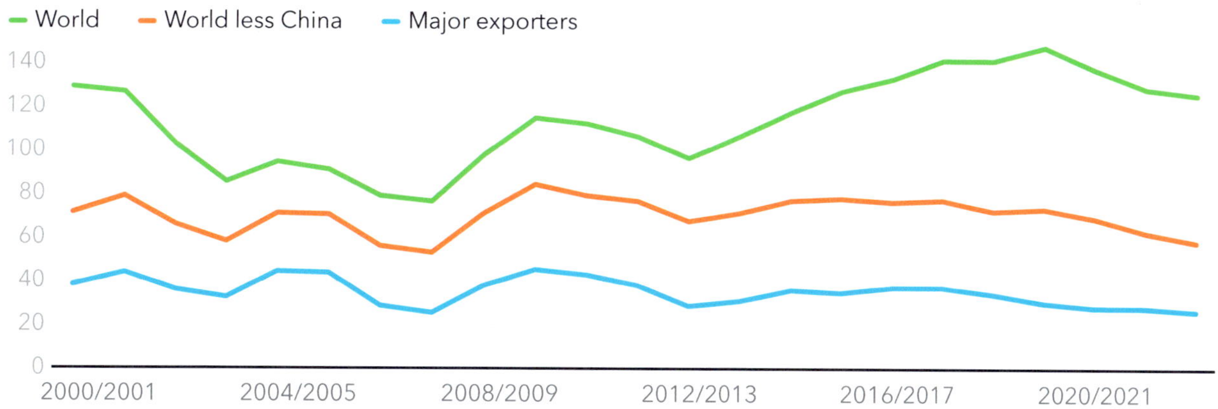

Note: Major wheat exporters include Argentina, Australia, Canada, European Union, Kazakhstan, Russia, Ukraine and the United States. These exporters accounted for over 85% of global wheat exports over 2020/21 to 2022/23.

Source: USDA/FAS PSD database

Over the historical range considered, stocks were at the lowest level for all three series in 2007/08, levels which most analysts consider close to pipeline levels.

More recently, the global measure has moved noticeably higher than the other two, mainly due to the increase in China's stocks. Global ending stocks peaked in 2019/20 at an estimated 147 days of use. Since then, ending stocks have fallen to an estimated 125 days of use for 2022/23, down 22 days since the recent peak, but far above 2007/08 levels (76 days of use).

Taking China out of the picture, the two other measures suggest stock levels are far tighter. If China is excluded from global totals, projected stock levels for 2022/23 are 58 days of use, the lowest level since 2007/08 (53 days). Similarly, considering only those stocks held by major exporting countries, ending stocks for 2022/23 are estimated at 26.3 days of use, also the lowest level since 2007/08 (25.4 days).

Again, it is important to reiterate that the stock measures considered here are imperfect at best and should not be interpreted literally as stocks currently on hand, but as an indication of relative tightness in the market.

Conclusions

The ongoing war in Ukraine has reduced wheat production in the country and disrupted markets worldwide. The relative tightness of global stocks suggests that price volatility will continue to remain high relative to the past 10 years. Going forward, rebuilding inventories of wheat and other key global crops would help to reduce both prices and price volatility. By the same token, tight stocks mean that an unforeseen production shortfall in a major wheat producing region would likely send prices sharply higher again (as in 2010/11 and 2012/13) and result in increased price volatility.

The paucity of reliable data on grain stocks continues to challenge market analyses. Despite efforts to improve data collection over the past 15 years, progress has been slow. National statistical offices, in collaboration with international partners and initiatives such as AMIS, should continue efforts to measure the size of reserves to help improve market monitoring and strengthen global food security.

11 Ukraine one year later: Impacts on global food security

Joseph Glauber

Originally published February 23, 2023.

When Russian troops invaded Ukraine one year ago, the war appeared to pose a grave threat to global food security. The conflict could hardly have come at a worse moment. Even prior to the war, global supplies of key staples were tight; ending stocks of wheat, maize, and soybeans were at their lowest levels for many years and prices were high compared to 2020 levels – lingering effects of the COVID-19 pandemic. The outbreak of war between two of the world's key agricultural producers jeopardized more than a third of world wheat trade, 17% of world maize trade, and almost 75% of world sunflower oil trade. Within a week of the invasion, prices of wheat futures had jumped almost 60%, while corn and soybean prices were up over 15% (Figure 1). Ukraine found its Black Sea ports effectively blockaded, sharply limiting its ability to export its 2021 crops, and the planting and harvest of its 2022 crops disrupted. The world faced the possibility of another food price crisis with potentially devastating consequences, following on the heels of the pandemic's global economic disruptions.

Figure 1: Wheat, maize, and soybean price evolution, US dollars per metric ton

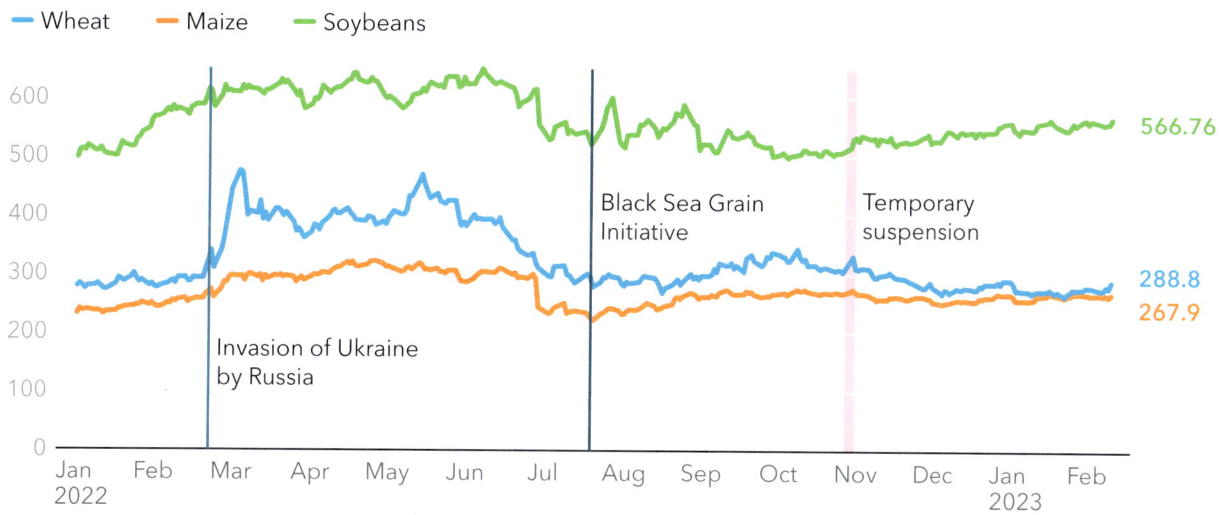

Note: Future prices (+30 days) quoted in Chicago
Source: CBOT

Yet as the war continued through 2022, international markets adjusted and adapted, and – while high prices and other problems persist – the worst-case scenarios for agricultural trade and food security were largely averted.

By August 2022 prices had fallen back to pre-war levels. Importers adjusted to market disruptions and found alternative suppliers. The partial re-opening of ports through the Black Sea Grain Initiative helped facilitate additional exports from Ukraine. In May, price volatility levels for wheat and maize reached the highest levels in more than 10 years, but by early February 2023 they had fallen back to within historical ranges. Lastly, the war precipitated an increase in global humanitarian efforts to help mitigate its impacts. For example, the World Food Programme reported that contributions from all sources in 2022 exceeded US$14.2 billion, up almost 48% from 2021 levels.

Now, with no end to the war in sight, what lessons should we take away from its food system disruptions to date? Should we no longer be concerned about the war's impact on markets and food security? This post reviews the experiences of the past year, and the uncertainties that remain about the future.

Markets have shown resiliency

Why did market prices fall? The invasion sparked high uncertainty in agricultural markets. First, Northern Hemisphere winter wheat crops had already been planted the previous fall, so producers were limited as to how much additional wheat could be planted. Second, the war appeared likely to disrupt Ukrainian and Russian exports, creating a major global supply problem.

Yet the worst fears of persistent global shortages did not come to pass (Figure 2a). Ukraine's wheat exports have fallen dramatically, by an estimated 5.3 million metric tons (a 28% decline) between 2021/22 and 2022/23, as have Argentina's exports, due to drought that sharply reduced its wheat harvest. But Russian exports are up strongly, due in part to a record wheat crop. That increase, combined with large crops in Canada and the European Union in 2022, equals a rise in projected combined exports exceeding 25 million tons – offsetting the export declines in Ukraine and other regions. Meanwhile, Australia's wheat exports continue to be robust, narrowly setting a record with an increase of 500,000 metric tons over last year, following back-to-back record export levels in 2020/21 and 2021/22.

Global maize exports show a slightly different picture (Figure 2b). Ukraine's maize exports in 2022/23 are down an estimated 4.5 million tons from 2021/22, but that is dwarfed by an estimated decline in US exports of almost 14 million tons – in part reflecting a reduced 2022 harvest and a strong dollar that has made US maize less competitive in world markets. EU maize exports are down as well, declining by 3.8 million tons. The declines are only partially offset by increased exports from Brazil and Argentina. Higher maize prices have reduced global demand and caused some importers to seek alternative feedstuffs such as broken rice. The U.S. Department of Agriculture projects a fall in global feed demand of almost 23 million metric tons (about 2.8%).

As crop production forecasts for wheat and maize brightened in the second half of 2022, and Russian wheat exports rebounded, market prices began to fall. In addition, markets remained remarkably resilient as importing countries shifted suppliers to meet demands. At the same time, global domestic wheat consumption in 2022/23 is projected to be largely unchanged from 2021/22 levels despite high prices, due in part to household safety net programs in some wheat-importing countries of the Middle East and North Africa.

While export restrictions exacerbated prices of grains, oilseeds, and other agricultural products in the months following the start of the war (as discussed in previous blog posts), by June many of the more binding measures had been removed or at least curtailed. That trend, combined with the partial reopening of Ukraine's Black Sea ports (where more than 22 million tons of grains and oilseeds have been shipped since August 1), has helped to add to world supplies.

Figure 2: Change in global wheat and maize exports, 2022/23 versus 2021/22, metric tons

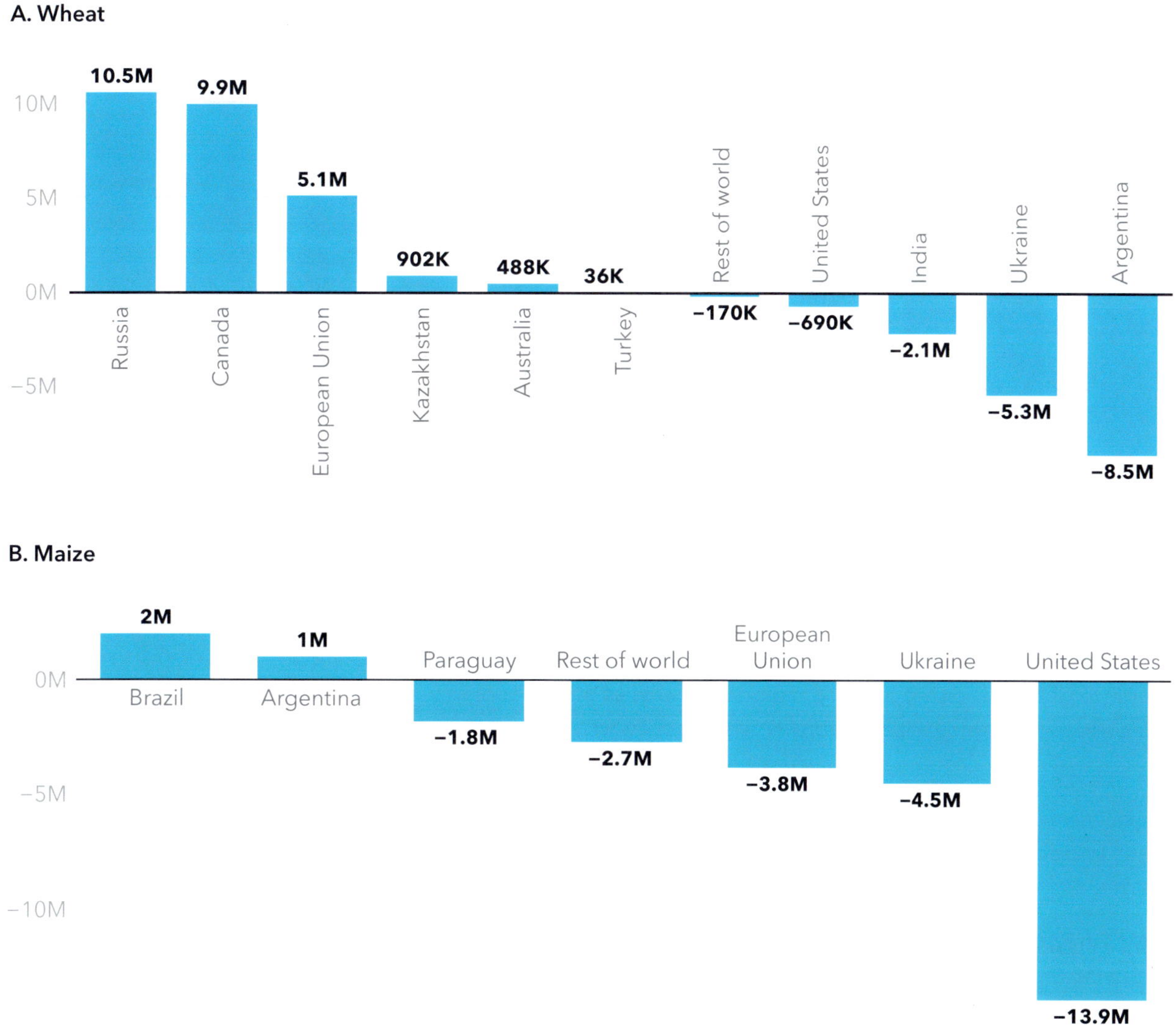

Source: USDA/FAS PSD database February 8, 2023

The impact of the war on Ukraine's agriculture to be felt even more sharply in 2023

But it is Ukraine itself that has felt the war's most direct and continuing impacts on agriculture. The war has already affected three successive crops:

2021 crops: Much of the previous year's harvests had not yet shipped when the invasion occurred, and could only be exported via slow and costly overland routes through the west. Though the development of so-called Solidarity Lanes (truck, rail, and barge routes through the EU) helped alleviate some of the backup, it was not until the Black Sea ports were partially reopened in August 2022 through the Black Sea Grain Initiative that agricultural exports began to approach more normal levels (Figure 3).

Figure 3: Ukraine's grain exports, metric tons

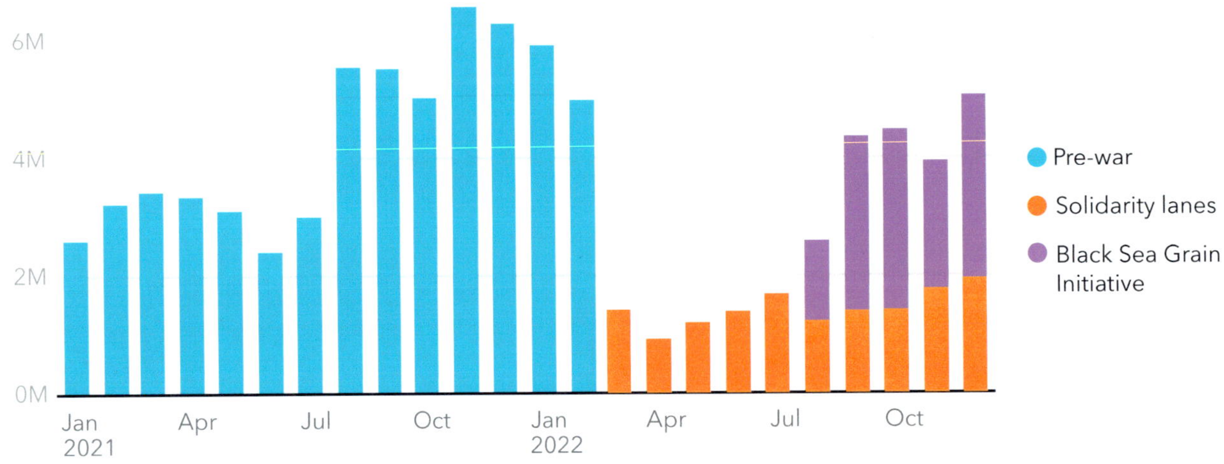

Note: Wheat, maize and barley exports. Solidarity lanes reflect grain not exported through Black Sea Grains Initiative.
Source: COMTRADE (through Oct) Ukraine Ministry of Agriculture (Nov-Dec)

2022 crops: In the meantime, producers struggled to plant, harvest, and market 2022 crops due to the direct impacts of the war itself and the indirect impacts of higher input costs such as energy and fertilizer prices. Production fell dramatically (Figure 4), with wheat and maize harvests both down 36%.

While efforts to facilitate export transportation have helped move more production to foreign markets, the associated costs have been high. Multi-modal (truck, train, barge) exports through the west are costly, while long inspection times and resulting demurrage charges have added significant costs to shipping through Black Sea ports. These costs have been largely absorbed by Ukrainian producers in the form of lower prices.

2023 crops: Those low producer prices, combined with high input costs, have discouraged planting for 2023 crops, and production is expected to drop even further.

Private estimates suggest Ukraine's grain harvest could fall by as much as 35-40 million tons in 2023, including 12-15 million tons of wheat and 15-17 million tons of maize. Wheat supplies for 2023/24 (2023 production plus year end stocks from the 2022/23 marketing year) are likely to be almost 30% below 2022/23 levels and 45% below 2021/22 levels (Figures 4). Expected maize supplies for Ukraine for 2023/24 could be 36% below this year's level and 53% below 2021/22 levels.

Impacts of fertilizer sanctions less clear

Sanctions imposed on Russia and Belarus in response to the invasion have disrupted energy markets and, relatedly, fertilizer supplies. While sanctions on Russia have explicitly exempted food and fertilizer, the country's fertilizer exports have lagged below pre-war levels, particularly due to the cutoff of anhydrous ammonia shipped via pipeline from Russian factories to port facilities in Odesa. Potash exports from Belarus have also dropped, as they are not exempt from current sanctions and are barred from export through EU ports on the Baltic Sea. The impacts are potentially significant since much of the world depends on Russia (and Belarus) for imports of nutrients.

Figure 4: Projected Ukrainian grain supplies, metric tons

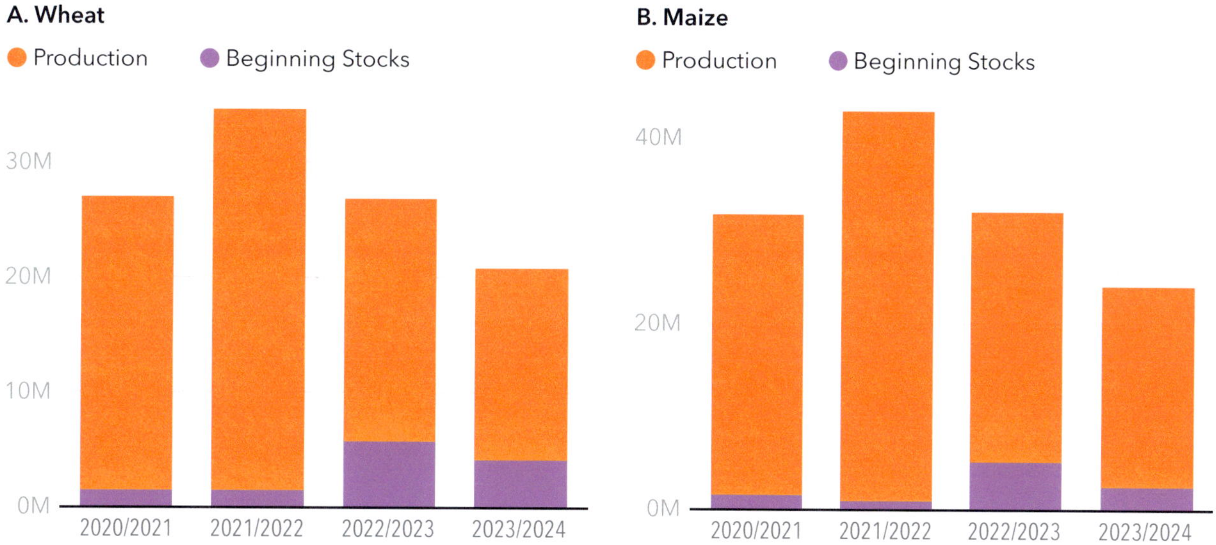

Note: 2023 crop estimates from Ukraine Ministry of Agriculture
Source: USDA/FAS PSD database March 8, 2023.

Yet energy and fertilizer prices have paralleled those of grains and oilseeds – peaking in the spring and then falling through the rest of 2022 as global production and markets adjusted to the shortfalls. The experience of major importers like Brazil and Indonesia suggests that as was the case with grain, markets have been very resilient, allowing them to find alternative suppliers (for example, Canada's potash exports). This has allowed many producers to meet their nutrient needs, albeit at persistently higher costs – despite the 35% to 50% drop in fertilizer prices since May 2022, prices remain 80% to 100% above 2020 levels. In addition, producers have shifted to less nitrogen-intensive crops like soybeans. For many producers in low-income countries, however, high fertilizer prices have made input purchases less affordable, potentially affecting yields.

Conclusions

With no end to the war in sight and threats of further escalation, uncertainty continues to hang over agricultural markets. Supplies are tight. Globally, estimated ending stocks for 2022/23 are at the lowest levels since 2007/08 for wheat and 2012/13 for maize and soybeans. Reduced plantings in Ukraine mean that the world will need to produce additional grains and oilseeds to help rebuild stocks and moderate price levels. Luck has played some role in events; in the year since the invasion, the combination of good weather and strong producer supply response has kept market prices from rebounding back to the high levels of early 2022. Tight stocks will mean increased price volatility, particularly during periods of uncertainty such as planting times and growing seasons in the Northern Hemisphere. In the meantime, uncertainty over events like the renewal of the current Black Sea Grain Initiative (currently scheduled to expire in March) will continue to roil markets.

Any of these uncertainties could lead to future market disruptions or other problems that bear continued close monitoring. The main lesson of 2022 is that global markets have generally worked well when not impeded by government actions like export restrictions. Hopefully that will be a lesson learned for countries as the world enters another year of global impacts from the war in Ukraine.

SECTION 2

POLICY RECOMMENDATIONS

12 Do no harm: Measured policy responses are key to addressing food security impacts of the Ukraine crisis

Joseph Glauber and David Laborde

Originally published April 12, 2022.

The Russian invasion of Ukraine is likely to have serious consequences for global food security. The UN's Food and Agriculture Organization has estimated that if there is a prolonged disruption in exports of wheat, fertilizer, and other items from Ukraine and Russia, the number of undernourished people worldwide could increase by 8 to 13 million people in 2022/23, with the most pronounced increases taking place in the Asia-Pacific region, followed by sub-Saharan Africa and the Middle East and North Africa.

Even before the crisis, agricultural commodity supplies were tight and market prices were at (nominal) record levels. Prices for most grains and oilseeds have risen sharply since the war began in late February (Figure 1). Russia and Ukraine together supply about 12% of global agricultural exports on a caloric basis, and over 30% of global wheat exports. The two countries command even higher shares in the wheat-dependent economies of the Middle East and North Africa, where wheat prices have risen over 30% compared to pre-war levels. Fertilizer and energy markets have seen rising prices and supply disruptions as well.

Figure 1: Commodity prices in constant US dollars (2022) per metric ton

Source: IFPRI based on CBOT quotation and US Bureau of Labor statistics

Given the potential food security impacts, particularly for the most vulnerable, many countries face difficult decisions. What approaches work best to limit impacts on the poor, maintain food availability, and cushion the blow of rising global prices – and what should be avoided?

In the short term, the primary message to policymakers should be "do no harm." Countries should refrain from enacting policies that could exacerbate price volatility and global shortages, and from taking precipitous actions that risk creating far more detrimental impacts in the long term. In particular, the price spikes of 2007/08 and 2010/11 and the more recent experience of the COVID-19 pandemic have shown that policies designed to insulate domestic markets from external disruptions can have serious adverse consequences for others, particularly the most vulnerable.

The most effective measures to combat food insecurity will be those that aim to keep trade in food and fertilizer products open and to mitigate the impacts of high food prices on the most vulnerable with targeted assistance and humanitarian aid.

1. Trade sanctions should exempt food and fertilizer

Since the invasion, sanctions against Russia have increased and added to existing measures taken after the annexation of Crimea in 2014. Trade sanctions are targeted to inflict economic pain on Russia in order to prompt Russia to reconsider its actions, but it is important that trade and financial sanctions continue to exempt food products and agricultural inputs like fertilizer. These products account for less than 5% of Russia's export earnings, but if sanctioned could have adverse impacts on global food security.

Russia's exports of natural gas, a key feedstock for producing nitrogenous fertilizers, is a more sensitive issue due to its role in Russia's economy (accounting for about 7% of its exports in 2019). While the European Union

Figure 2: Food security do's and don'ts, 2022

DO

- Remove biofuel subsidies and mandates
- Target social safety nets to the most needy
- Boost funding to WFP and other humanitarian programs to reflect rising food needs and operating costs
- Allow market prices to guide producer and consumer decisions

DON'T

- Apply sanctions that obstruct food and fertilizer trade
- Implement export restrictions
- Engage in panic buying
- Target subsidies to specific crops to "engineer" production outcomes
- Cancel environmental initiatives (including land set asides) without weighing long-term costs
- Promote self-sufficiency policy and autarky strategy

Source: IFPRI

continues its purchase from Russia, any disruption in supply will have even larger consequences on European fertilizer production and farm input availability than current high energy prices. Beyond European borders, the capacity to produce additional urea or ammonia is larger and mitigation options more available. Policymakers should consider the indirect effects of Western sanctions, and the impacts on third parties, like Türkiye, that play a critical role in processing grains imported from Russia and re-export final or semi-transformed products to other low- or middle-income countries. Appropriate legal and technical support should be provided to partner countries and small and medium enterprises (SMEs) to navigate new regulations.

2. Countries should refrain from implementing export bans and restrictions

During the food price spikes of 2007/08 and 2010/11 and again during the first months of the COVID-19 pandemic, a number of countries restricted exports in order to ensure adequate supplies at home. But export restrictions can also drive global prices even higher, making it even more difficult for net food importing countries to purchase food. Moreover, export bans tend to be contagious, as other exporting countries follow suit and implement their own bans. The way the global community handled this issue in early 2020 should be an example to follow now: Most countries refrained from implementing bans, and those that were implemented were relatively short-lived. However, the current trend is worrying, as shown in the IFPRI Export Restriction Tracker and discussed here.

3. Countries should avoid hoarding and panic buying

Panic buying can disrupt the orderly marketing of commodities and drive prices up in the short run. Supply hoarding – importing and storing many months of grain – can exacerbate price volatility and potentially be costly as prices fall over time as more supplies become available. With improved market information provided by international efforts such as the Agricultural Market Information System (AMIS), countries can make more informed purchasing decisions and hopefully avoid the panic buying that occurred during earlier price spikes.

4. Countries should target food subsidies toward the most needy

While the world will not run out of wheat, prices are likely to remain high for some time, and could go higher if harvest problems cause further drawdowns in global stocks. High food prices are difficult to absorb in poor households with limited resources and that consume a limited range of foodstuffs. To avoid an increase in food insecurity, and as was seen in 2007/08 and 2010/11, food riots and increased political instability, countries have provided subsidies to keep food costs low. Accurate targeting is crucial to ensure that subsidies go to the truly needy and not to more prosperous households that can absorb increased food costs, or households that can readily switch to lower-cost alternative foods. Too often, existing schemes (for example, in Egypt) are poorly targeted and waste significant resources, especially when rising prices increase the cost of running domestic food aid programs.

5. Countries should provide humanitarian aid through programs such as WFP

Some countries remain heavily dependent on humanitarian aid to meet food security needs. Higher commodity prices mean humanitarian aid agencies like the World Food Program (WFP) will have to spend far more than previously budgeted to meet those needs. Countries in the position to do so should ensure that WFP and other organizations are adequately funded. To the extent possible, assistance should not be tied to national export interests – organizations like WFP operate most efficiently when they are able to source food from the lowest cost suppliers.

6. Countries should suspend biofuel mandates and subsidies

Global biofuel production is at record levels. In the United States, for example, over 35% of the 2021 maize crop will go to ethanol production, and over 40% of US soybean oil production goes to biodiesel production. Globally, 31 million tons of vegetable oils are converted into biodiesel (Figure 3), about 15% of global production – the equivalent amount of calories required to feed more than 320 million people per year. With vegetable oil prices skyrocketing (they are now around $1,600 per ton, after hitting $2,000 one week after the invasion), now is not the time to be encouraging the conversion of food crops to energy through artificial policy incentives or mandatory blending targets. While beneficial to farmers and landlords, such policies come at the expense of those who can least afford it, hungry people around the world.

Figure 3: Consumption of vegetable oils by the biodiesel industry, metric tons

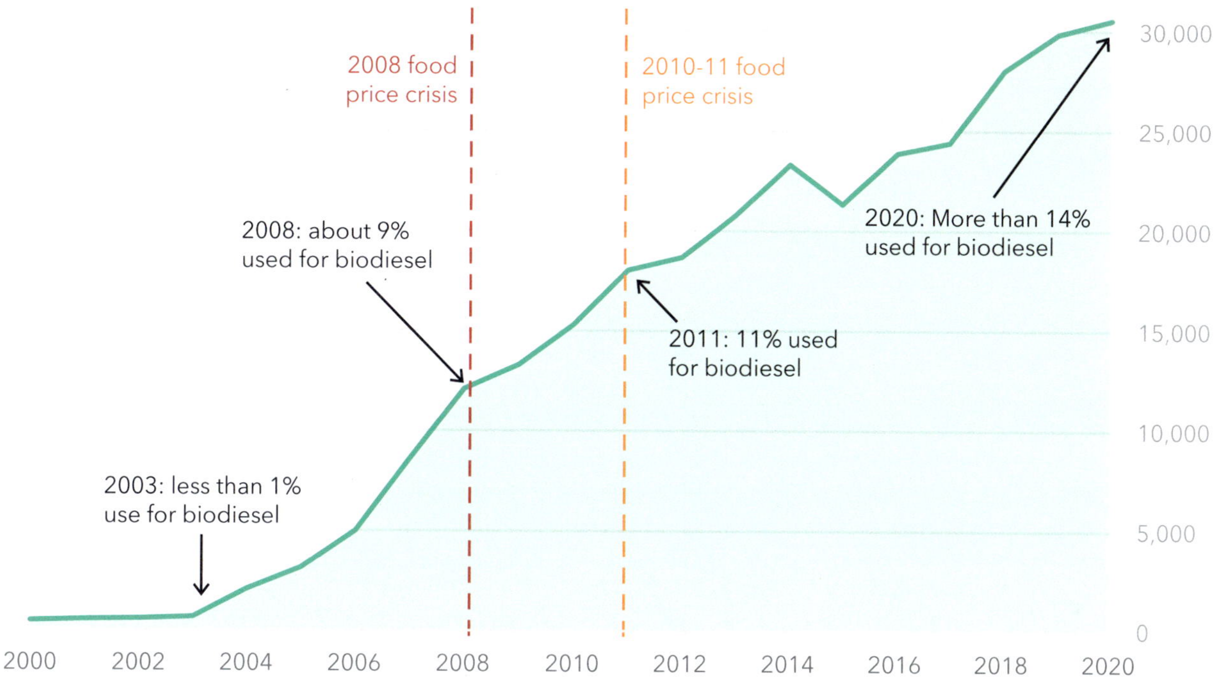

Source: OECD-FAO

Domestic policies to promote increased food supplies: Beware unintended consequences

As it is increasingly likely that supplies of key commodities from Ukraine and Russia will be severely disrupted for 2022, countries are seeking ways to boost global supplies to compensate. Policymakers should take care to ensure that any emergency measures they take do not lead to additional problems in the short or long run; like previous crises, the current one will pass and agricultural prices will retreat to their longer-term dynamics. Many emergency policy decisions have long-term effects, for example, damaging biodiversity or contributing to climate impacts by releasing carbon or methane. In addition, policy U-turns reduce the credibility and predictability of policies, thus undermining their effectiveness and reducing the collaboration of the private sector.

7. Countries should avoid introducing new subsidies to encourage production of one crop over another

Subsidies distort production decisions and obscure market signals, potentially having unintended consequences and costs for others. Moreover, once introduced, history has shown that subsidies are very difficult to remove, prolonging their impacts on global supply and demand.

8. Releasing land from fallow and long-term set asides will not likely help in the short run but new enrollments should be delayed.

The recent announcement by the European Union to allow fallow land to come into production and similar calls in the United States to release land from the Conservation Reserve Program will likely have little immediate impact on global food supplies. Much of the land has been out of production for many years and is of marginal productivity. It is unlikely that much land could be brought into cultivation in time for the current crop year and many of the environmental benefits of long-term set-asides, such as carbon sequestration and biodiesel production, could be lost. Record-high prices will likely dissuade producers from entering new lands into such programs; countries may wish to delay enrollments until the end of 2022, when more will be known about the impacts of the war.

9. Safer alone? The trap of promoting food self-sufficiency

When world markets are in turmoil, a traditional response, happily supported by domestic lobbies, is to relocate production within national borders. Subsidies and tariffs are the preferred policy tools, and building large domestic reserves a principal aim. Unfortunately, this response is factually and conceptually fallacious. World markets are more stable than domestic markets, because they are larger and more diversified. The global farming community is more flexible and able to react to shocks more rapidly than any individual country. For example, the world has two hemispheres and planting decisions to take place year round; by contrast, in a given country the adjustment period typically takes 12 months on average. Modern global value chains operate with smaller food losses than large public stock programs.

Self-sufficiency is also prohibitively expensive for most net food importing countries, many of which are among the poorest in the world. The world will become less, not more food secure, if countries resort to self-sufficiency policies that expose them to production uncertainties caused by drought, floods, pests, and animal or plant diseases – all increasingly frequent with climate change. Policies should not segment markets, but rather aim at creating more opportunities for a larger number of countries – helping global markets become more diversified and inclusive.

Conclusions

Russia's invasion of Ukraine will likely mean that high agricultural prices continue through 2022 and into 2023. Unfortunately, most policy responses, however well-meaning, are unlikely to provide much short-term relief, and could exacerbate global price increases and instability. The best approach may be the simplest: Allow markets to work by removing distortions, and support the most vulnerable countries and households via social safety nets, and where most needed, through humanitarian assistance. The current crisis presents a number of immediate humanitarian challenges (particularly for those in war-torn Ukraine), but over time, markets should eventually return to more normal patterns. In the meantime, as countries take action to respond to high prices and other fallout from the crisis, they should take care to avoid exacerbating the impacts for others.

13 Short-term policy considerations to respond to Russia-Ukraine crisis disruptions in fertilizer availability and affordability

Charlotte Hebebrand and David Laborde

Originally published June 8, 2022.

As discussed in an earlier post, fertilizer prices have increased drastically, up 230% between May 2020 and May 2022. Now – driven upward by supply disruptions stemming from the Russia-Ukraine conflict – they are nearing the peak reached in August 2008 during the last major food price crisis.

Spiking fertilizer prices are likely to have major impacts on agriculture and food production – and thus, food security – around the world, as farmers struggle to pay for a key input and face potential supply disruptions, and governments look for ways to cushion the economic blow.

Figure 1: Synthetic fertilizer application rates and cereal yields, 2019

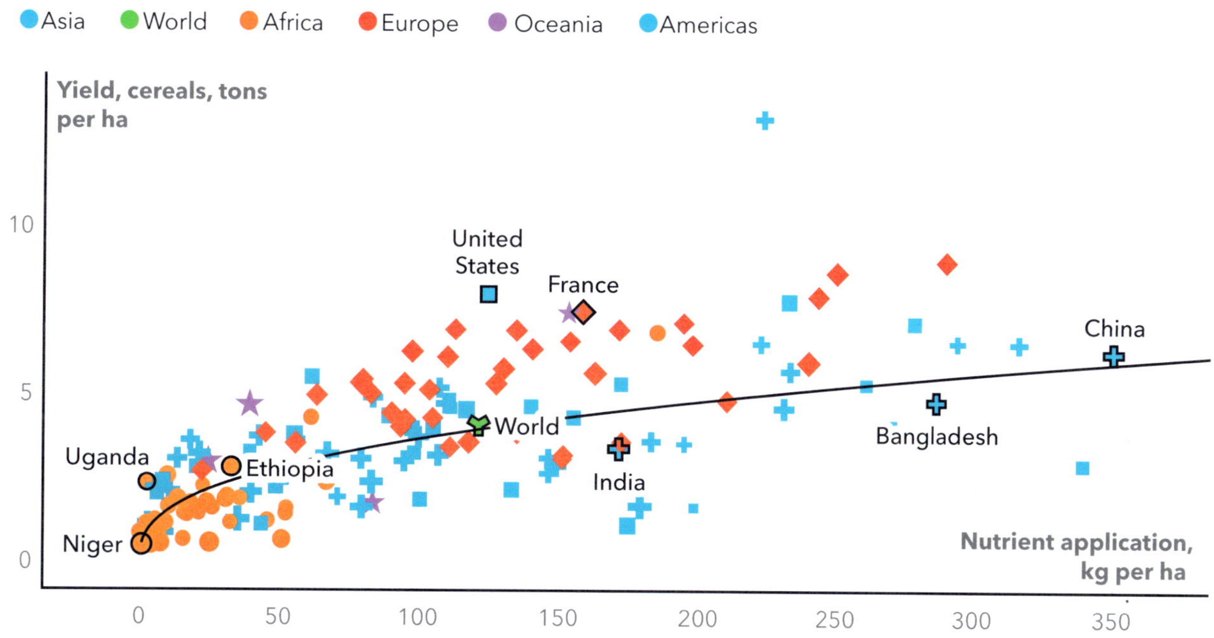

Source: FAOSTAT

How should policymakers respond to this complex emerging situation in fertilizer markets – in particular, to reduce impacts on food security?

In this post, we set forth some considerations for policymakers and other stakeholders as they put short-term solutions into place to address fertilizer availability and affordability concerns. A later post will examine the medium- to longer-term considerations.

1. Maintain global trade to ensure continuing availability of fertilizers

Removing existing export restrictions, and foregoing new ones, is an important priority now. Export bans, stringent inspections and licensing processes, and sanctions that limit availability contribute to supply disruptions and push market prices higher. Currently, such restrictions impact 20% of global fertilizer trade and threaten more than 50% of fertilizer supply for 24 countries (Figure 2). Governments should also designate fertilizers as essential goods to ensure that they can continue to be traded and prioritized in logistical channels.

Figure 2: Share of fertilizers impacted by exporter restrictions

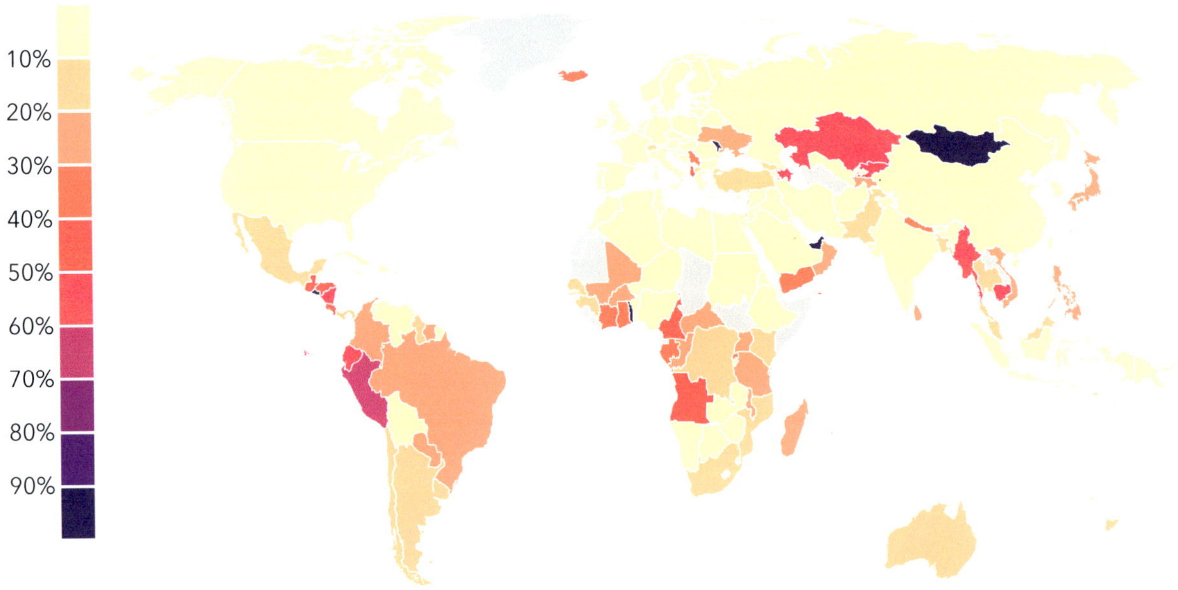

Source: IFPRI

The continuing impacts of economic sanctions on Russia, one of the world's largest fertilizer producers and exporters, along with Belarus, also a key supplier, should be monitored, with an eye toward understanding their implications for fertilizer producers, traders, and buyers around the world. The European Union and United States have not imposed sanctions on agricultural products from Russia or Belarus, but they have sanctions in place targeting the main Belarus companies involved in potash production and trade. There is also concern that financial sanctions imposed by the EU and United States are hindering trade of Russian and Belarusian food and fertilizers to third countries by impacting logistical and financial operations for many businesses with connections to the agricultural economy. When we combine both export restrictions and export disruptions, 32 countries in the world have one-third or more of their fertilizer use at risk.

Governments should provide specific financial and technical solutions to mitigate potential impacts of sanctions on shippers, traders, and others along value chains, so that grain and fertilizers from these two countries can be exported. For instance, "comfort letters" should be issued by major countries, following the example of the United States, to ease the anxiety of financial and insurance companies. Legal helpdesks should be set up to help medium-sized firms involved in fertilizer operations navigate the sanctions, and/or special credit lines guaranteed by public financial institutions for these companies when their private banks decide to suspend operations with Russian counterparts.

An option for governments seeking to secure short-term supplies of fertilizer and/or food is employing a form of "barter trade," that is, bilateral deals allowing countries to swap fertilizer shipments for agricultural commodities. Indeed, several fertilizer exporting countries are also traditional grain importers – in particular, some countries in the Middle East and North Africa region.

The goal is not to make such government-to-government trade arrangements a general policy prescription, but to propose a short-term alternative to imposing export restrictions. Such deals could also reduce the temptation of some countries with unsuitable climates or water resources to achieve costly (economically as well as environmentally) food self-sufficiency.

2. Help all farmers increase their fertilizer use efficiency

Just as it's important to pursue efforts to reduce food loss and waste in the current climate of high agricultural commodity prices, increasing fertilizer use efficiency (the amount of fertilizer applied that is taken up by crops at harvest) is a crucial priority in a context where supply will remain constrained this year.

Extension services can provide the know-how and tools to facilitate more efficient, crop- and site-specific, and integrated nutrient management, which can help to reduce supply pressures while also improving efficiency and minimizing nutrient losses to the environment.

The priority should be educating farmers to implement the "4Rs" (applying the right nutrient source, at the right rate, at the right time, in the right place). To be effective, the 4R principles should be promoted at scale by extension services and private sector actors, as many of the world's farmers still lack the proper tools, knowledge, or incentives to undertake site-specific nutrition management (SSNM).

SSNM was a concept first developed in the 1990s by CGIAR researchers (for rice) and then modernized through tools also developed by CGIAR, such as the International Rice Research Institute's (IRRI) rice crop manager and the International Institute for Tropical Agriculture's (IITA) AKILIMO agronomic advisory service for smallholder farmers.

Farmers should also be encouraged to diversify fertilizer sources. A short-term priority here is recycling more livestock manure and considering manure as a resource rather than waste. Proper nutrient management and following the 4Rs is just as important for manure application as it is for mineral fertilizers to ensure efficiencies and reduce losses to the environment.

3. Support the most vulnerable

Governments will be under pressure to provide relief to farmers facing high fertilizer prices via subsidies and other forms of support. Such assistance should be limited to the most vulnerable farmers in low- and middle-income countries (LMICs) in Africa, Asia, and Latin America.

Direct fertilizer subsidies should be temporary, targeted, encourage efficient and balanced plant nutrition, work in tandem with rather than compete with small and medium enterprises (SMEs) in the value chain, and include proper tracking to avoid diversion and smuggling. Regional coordination and harmonized policy responses, also in terms of subsidy regimes, are required to ensure that fertilizers can easily cross borders. Similarly, fertilizer price controls should be avoided since they usually negatively impact input dealers and create black markets.

Additional financial support may also be necessary for some farmers facing high input costs that they cannot recoup because of comparatively low market prices for their crop which, for example, is the current situation of rice producers. In addition, social protection measures such as cash transfers or food distribution vouchers should be provided to farmers unable to access fertilizers this year to protect their own food security.

Governments can also provide incentives to smallholder farmers to prioritize the production of food crops over cash crops, and design legal and financial solutions for companies involved in cash crop businesses, allowing greater flexibility to contract farmers. It is more efficient to compensate cash crop producers for their crops than to subsidize fertilizers for cash crops.

4. Provide financial support for SMEs and fiscally stressed governments

Amid uncertain supplies and volatile prices, LMIC governments should work to ensure that SMEs in the fertilizer value chain have continued access to financing to ensure the ongoing blending and distribution of fertilizers.

The international community will also be called upon to assist LMICs facing high fertilizer prices and supply issues. When such efforts involve financing subsidies, care should be taken that they do not further fuel inflation or create unfair competition between LMICs in need of fertilizers. LMICs in a precarious financial situation due to the economic shocks created by COVID-19 and/or facing acute macroeconomic constraints, such as Sri Lanka, should receive priority support, and such efforts should include a strong grant element to avoid exacerbating debt problems.

5. Ramp up supply in the short run

The best remedy for high prices is increased supply. Typically, brownfield and greenfield expansions of fertilizer production imply huge costs and take years to complete, but there is some new capacity coming online in the short to medium run.

The nitrogen (N) market is severely impacted by the current crisis, considering that Russia represents 15% of this market globally, export restrictions are impacting 22% of global N trade, and prices for natural gas – central to the N production process – will likely continue to cause instability in European production. But new capacity is coming online in Nigeria and in India, and there are longstanding plans to increase production capacity for urea, a nitrogen fertilizer; five new plants are now at various stages of commissioning. India is currently a large importer of urea, so becoming self-sufficient would free up urea for other markets, though it is still unclear whether this goal is feasible considering the reliance on costly regassified liquified natural gas (LNG).

The phosphate (P) sector also faces uncertainties regarding Russian exports (14% of global trade), and a further 25% of traded P may be impacted unless China, the top global producer, lifts its export restrictions. Other P-producing countries may be positioned to expand production. In March, for example, Morocco – the world's second-largest phosphate producer – signaled its intention to increase its output by 1 million metric tons in 2022 from the current level of 4 million tons, and to add an additional 3 million tons in 2023.

While some exports of potash (K) from Russia and Belarus may continue, they will likely fall far short of their 2019 combined 37% global market share. With its 34% share of global K trade, attention clearly shifts to Canada. With the price outlook for K improved, an international mining company announced plans for a major K production mine and site, while in March 2022 another major Canadian potash producer announced its intention to ramp up K production by an additional 1 million tons by the second half of the year – an 8% increase of the total country output.

More capacity could still come online in Canada, and smaller producers such as Germany, Israel, and Jordan could also expand their production in the short term. In the medium run, other players could emerge. A debate is underway in Brazil about potentially mining unused potash resources; however, this could come with very high social and environmental costs. A plan to process phosphate rock imported from Morocco might be a better way to address Brazilian concerns about access to P. But neither approach is feasible in the short run. Care should be taken to ensure that long-term sustainability is not sacrificed to address an immediate crisis.

6. Improve market intelligence and enhance transparency

In this time of rising food insecurity and market volatility, it is vitally important to collect and provide accurate and timely information on fertilizer supply, including on government or private sector reserves. Such data can help governments, producers, and consumers avoid making misinformed decisions, including panic buying or hoarding, and avert price manipulation by actors wishing to take advantage of the crisis.

Figure 3: Short-term actions to address the fertilizer crisis

- Make economic sanctions harmless for fertilizer trade with "comfort letters" and open port permits.
- Remove export restrictions and taxes on fertilizers.
- Fertilizer subsidies should be temporary, targeted, and not compete with private sector distribution.
- Specific financial facilities should be provided to countries that face major macroeconomic or exchange rate crises.
- Public spending in LMICs should focus on reducing trade costs impacting the access and distribution of fertilizers and support access to finance for SMEs.
- Identify smallholder groups that will face fertilizer shortages in 2022 and 2023, and prepare safety nets to compensate for expanded production losses.
- Scale up extension service efforts to optimize fertilizer use and support all farmers with improved knowledge.

The G20 Agricultural Market Information System (AMIS), convened by the UN's Food and Agriculture Organization (FAO) and the Organisation for Economic Co-operation and Development (OECD) should be expanded to include closer monitoring of fertilizer markets. At the same time, initiatives like IFPRI's Fertilizer Dashboard, along with the data gathering and monitoring efforts of the International Fertilizer Association and AfricaFertilizer.org, should continue to receive support and be integrated in the wider system of agricultural statistics centered around FAO and FAOSTAT, and also provide indicators and tracking for sustainability challenges.

Conclusion

These recommendations offer a variety of ways to blunt the immediate impacts and uncertainty of high fertilizer prices and uncertain supplies for those who are most vulnerable – farmers in LMICs – and to sustain agricultural production and food security. Governments and organizations should act quickly and judiciously to support farmers, while steering clear of the pitfalls of trade restrictions and high subsidies. We summarize immediate actions required in Figure 3, on which the Group of Seven (G7) and the Group of 20 (G20) countries could and should act now. In our next post, we will examine some longer-term considerations for fertilizer markets and policies.

14 The response to the global food crisis must address the needs of women and girls

Elizabeth Bryan and Claudia Ringler

Originally published August 31, 2022.

Development agencies are pouring in billions of dollars to address the global food crisis exacerbated by Russia's war on Ukraine. The World Bank, the G7, the European Union's Team Europe, and the United States have collectively pledged more than $40 billion to avert food and humanitarian crises. Yet this much-needed assistance carries its own risks. Without a gender lens, the proposed measures will fail to meet the specific needs of women and girls and might worsen existing gender inequalities. The current crisis presents an opportunity to design gender-responsive programs that buffer women and girls against hardships in the short term and tackle entrenched gender inequalities while building resilience to future shocks over the long term.

It is a perilous moment for women and girls around the world. The COVID-19 pandemic already pushed an estimated additional 47 million girls and women into extreme poverty, reversing decades of progress. And in 2021, at least 150 million more women than men were experiencing food insecurity – with the gender gap continuing to grow. Crises pose particular risks to girls, leading to them to drop out of school at higher rates, and increasing the incidence of gender-based violence, including early or forced marriage and economic or sexual exploitation.

The war in Ukraine has led to price spikes of key staple foods, fuel, and agricultural inputs around the world, posing special risks for low-income countries with dependence on food and fertilizer imports. Rising food insecurity and malnutrition pose the greatest risks for children and pregnant and lactating women, who may suffer repercussions to health and productivity that can affect future generations. The rising cost of fuel and farm inputs threatens to widen the gender gap in agricultural productivity and other economic opportunities, leading to further reductions in women's income and household food security.

The strain on government budgets will likely shift spending away from social programs, limiting the provision and expansion of social protection measures needed to reach women and children. Other responses to the crisis, such as the provision of fertilizer subsidies or vouchers, are more likely to reach male heads of household as gender disparities in access to resources or services like financial accounts and mobile internet are ignored. Moreover, many of the practices proposed to address the crisis, such as the use of agroecological practices, may add to the already high labor burdens of women and girls.

Alarmed that the crisis response would ignore gendered impacts, IFPRI's Gender, Climate Change, and Nutrition Integration Initiative (GCAN), with support from the US Agency for International Development (USAID), convened a roundtable of concerned donors, academics, and practitioners. The group identified eight actions, now published in a Nature Comment, that chart a course for concrete, gender-responsive actions to support the needs of women and girls:

Gather evidence. Gender-responsive approaches require data on food crises broken down by sex that are timely, accessible, and localized. Monitoring longer-term effects on women's empowerment and well-being is also essential. Without such data, gender-blind investments are difficult to avoid.

Extend anti-violence programs. Crises raise the risk of gender-based violence both inside and outside the home, including early and forced marriage as a coping mechanism. More services for those who experience violence are needed, as are prevention measures. These include cash transfer programs, measures designed to keep girls in school, and programs that address harmful attitudes and norms by working with men and boys.

Expand social protection. Evidence abounds that social protection programs can protect women and children during food crises. It should be a priority to expand these programs to women who are not formally in the labor force and those whose businesses were forced to close or are struggling.

Increase opportunities for decent work. Women need support to run businesses, including farming. Such support includes childcare, business development centers, and access to finance, as well as targeted procurement of goods and services from women farmers and entrepreneurs.

Support women's groups and networks. Existing social networks and women's groups are essential to direct resources to women and girls. They can also boost resilience by providing a platform for collective action, sharing labor and childcare responsibilities, organizing transport, accessing credit and savings, and disseminating information.

Tailor financial services to women. Waiving women's access barriers, such as registration fees or collateral requirements, while simultaneously increasing financial literacy, should be part of any program fighting food insecurity.

Broaden access to information. To counteract current agriculture and food insecurity challenges, women need information on less labor- and input-intensive farming practices and how to safely store food to prevent contamination and minimize food waste. Increasing their ownership of mobile phones and training on how to use them is essential to access extension messages and market information increasingly delivered through digital platforms.

Promote women's leadership. Women leaders and women's organizations need equal and meaningful participation in decision-making spaces at multiple scales to develop more inclusive processes, design gender-responsive interventions, coordinate implementation across sectors, and monitor and evaluate the gendered impacts of the crisis response. Only if women's voices are formally represented in crisis planning and response will programs be more equitable and effective.

The current global food crisis is not the last crisis the world will face, but it should be the last one in which women and girls carry this grossly unequal burden. Now is the time to build more resilient and equitable food systems that create more opportunities for women and girls, leading to greater gender equality and greater prosperity for all.

15 The global food price crisis threatens to cause a global nutrition crisis: New evidence from 1.27 million young children on the effects of inflation

Derek Headey and Marie Ruel

Originally published December 14, 2022.

Frequent food crises with spiking prices have become the new normal in the 21st century, bringing urgency to the task of understanding their nutritional impacts on poor and food insecure populations. In a new analysis of 1.27 million children in 44 low- and middle-income countries (LMICs), we show that exposure to food inflation in the womb and first years of life is associated with greater risks of child wasting in the short run and stunting in the long run. Food inflation also poses larger wasting risks for children of poor and landless rural households, exacerbating existing inequalities.

These results show the urgency of policies focused on improving maternal and young child nutrition, and the importance of broadening actions to reduce food price volatility and boost access to nutritious foods.

The age of volatility: Food prices in the 21st century

Food prices have become extremely volatile over the past two decades. We are now in the midst of the third international price spike in 15 years, as food prices surged upward in late 2021 due to the tailwinds of the COVID-19 pandemic, before skyrocketing in the wake of Russia's invasion of Ukraine. By March 2022, the UN's Food and Agriculture Organization (FAO) international Food Price Index had reached an all-time high, 116% greater than its 2000 value – higher even than the peaks reached in the 2007/08 crisis. More recently, international prices have fallen but domestic food inflation in LMICs remains high by historical standards, and exceptionally so in countries hit by economic mismanagement, conflict, or poor governance. For example, food prices in Sri Lanka had risen 100% by the third quarter of 2022 compared to 12 months earlier, but also soared in Pakistan, Laos, and Ethiopia.

Increases in international food prices are clearly a threat to the welfare of the poor, especially after the severe impacts of COVID-19 on poverty and malnutrition. IFPRI simulation models suggest that rises in food, fuel, and fertilizer prices are associated with poverty increases in the short run, even if rural economies can benefit from higher prices in the longer run.

Little is known, however, about the impacts of food inflation on child nutrition in the short or long run. Optimal maternal nutrition during pregnancy and child feeding practices in early life play a critical role in stimulating development and protecting health and nutrition at all stages of the life course. So as food prices rise and households' access to nutritious foods and healthy diets declines, a major concern is that the nutrition of pregnant and lactating women and their young infants, whose nutrient requirements are high, may suffer the most.

Figure 1: A century of volatility? Trends in the real Food Price Index, Jan. 2000–Aug. 2022

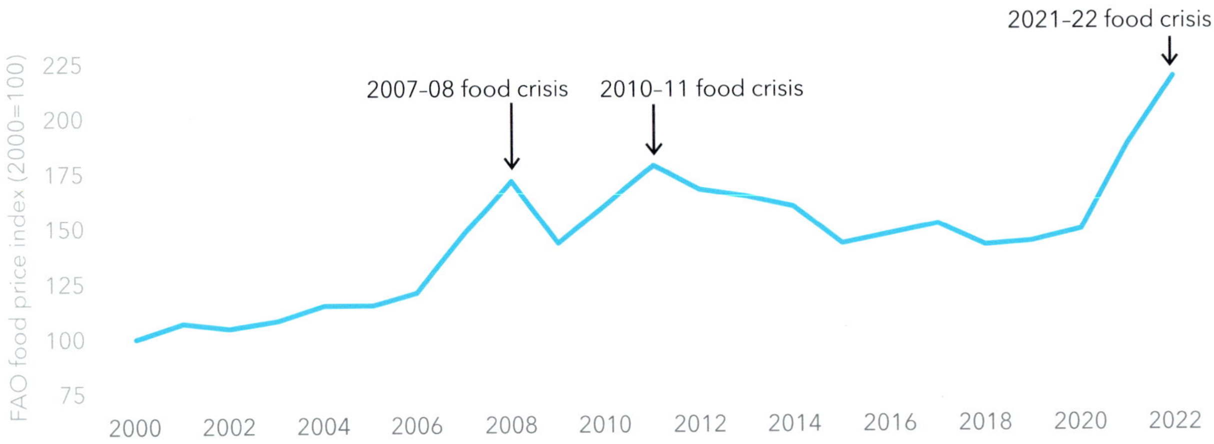

Source: Data are the author's construction from the real FAO international food price index. The 2022 value refers to the average of January-August 2022.

While one study from Mozambique examined the impact of food inflation on nutrition at the country level, our study is the first to look at this question on a "global" level. It uses nutritional data from 130 Demographic Health Surveys (DHS) conducted over 2000-2021 in 44 LMICs, which we link to a monthly index of national changes in the ratio of food prices to overall consumer prices – i.e., "real" food price changes.

For wasting – a measure of short-term or acute malnutrition – we test the effects of food price increases in the three months prior to wasting measurement. Those predicted risks are shown in Figure 2 for all wasting (in black) and severe wasting (in blue), by child age. Food inflation is associated with higher risks for wasting and severe wasting and the effect size is large and generally robust across ages. Across all children 0–59 months of age, a 5% increase in the real price of food is associated with a 9% higher risk of wasting and a 14% higher risk of severe wasting. In separate tests, we find that infants 0–5 months of age are also at increased risk of wasting, suggesting a maternal nutrition pathway during pregnancy. However, food inflation was also associated with poor dietary diversity among children 6-23 months, suggesting an additional postnatal pathway.

Which are the most vulnerable groups?

Poverty studies tell us that poor, rural and landless households are typically more vulnerable to inflation shocks, while nutrition-focused studies often find that male infants are more vulnerable to shocks than females. Here too we find that food inflation predicts larger wasting risks for male children, rural children, and children from asset-poor and landless households (which lack the option of growing their own food in the face of rising food prices in markets) (Figure 3).

The long-run nutrition costs of short-run economic shocks

Even brief nutritional insults can have long-term consequences on child growth and development, especially insults that occur during the first 1,000 days of life when the biological foundations for optimal nutrition, health, and development are established. Our study found that price increases during pregnancy and the first year of life were associated with a greater subsequent risk of stunting at age 3-5 years, with stronger effects for severe stunting.

Figure 2: Estimated impacts of a 5% increase in the real Food Price Index on wasting by child age

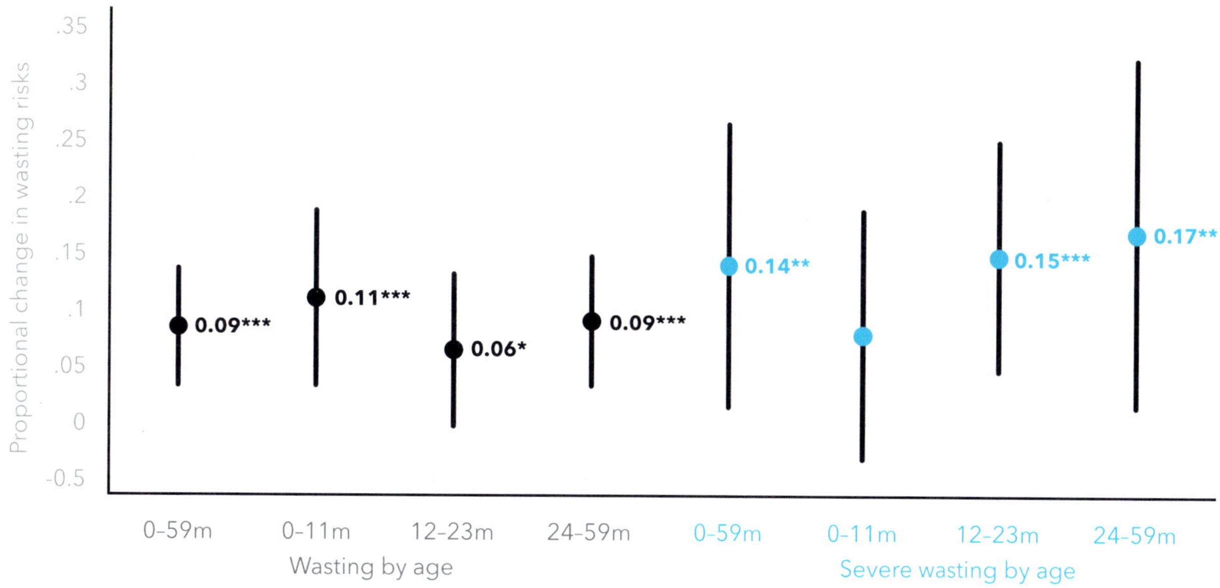

Note: Points refer to coefficient estimates from a linear probability model while lines refer to 95% confidence intervals. The full sample of children 0-59 months includes 1.271 million children in 44 LMICs. Wasting refers to WHZ<-2 and serve wasting to WHZ<-3, where WHZ are weight-for-height z scores/

Figure 3: Predicted impacts of food inflation on wasting risks, by gender, location, poverty status, and land ownership

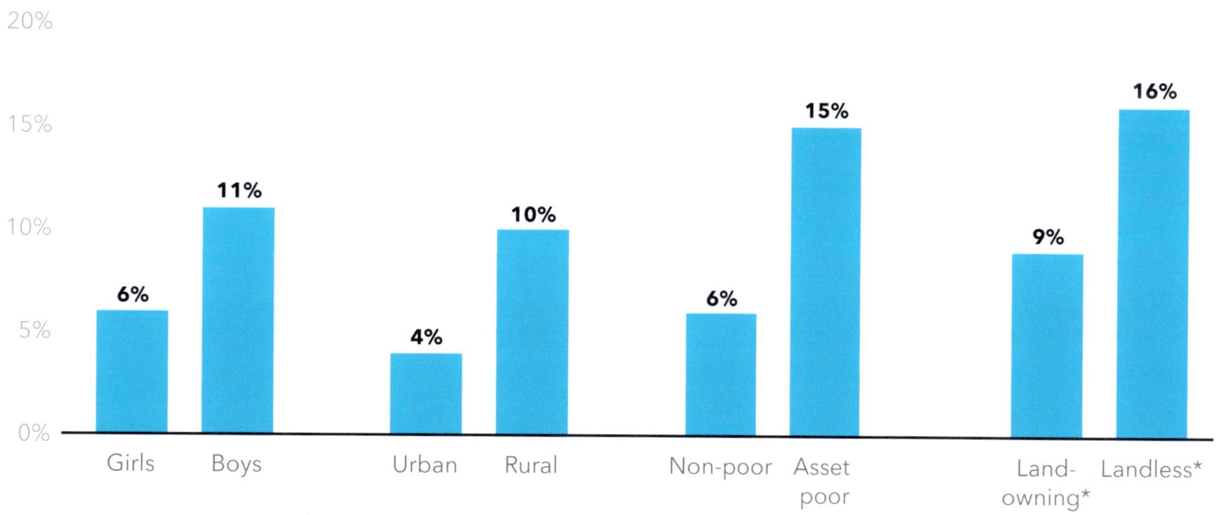

Note: These are coefficients derived from linear probability regression models that interact food inflation with gender, location, poverty status, and land ownership (*in rural areas only). All interaction terms were statistically significant at the 5% level or higher. The full sample included 1.271 million children in 44 LMICs. Wasting refers to WHZ<-2.

Policies and programs to protect young children and mothers

This study has important implications. The impacts of food inflation strike very early – in the first 1,000 days of life, during pregnancy and infancy – and hit the rural poor and landless especially hard. These results demonstrate the urgent need to strengthen women's nutrition and health before and during pregnancy to ensure optimal prenatal health and nutrition support for their own nutritional benefit and that of their young child.

Maternal and child food or cash transfers (possibly with health/nutrition conditionalities) could provide protection throughout the first 1,000 days and beyond, especially if they are targeted at the poorest groups and combined with nutrition- and health-focused social behavior change communication, and adjusted for inflation. Multidimensional early warning systems also warrant further investment in an era of greater food price volatility and more extreme weather events, as do programs designed to prevent and manage severe acute malnutrition. Finally, food policies should aim to achieve greater stability in food prices and improve the affordability of nutritious foods. Many suggested policy responses to the 2007/08 crisis are still relevant today; so too is more contemporary advice on repurposing agricultural policies to make healthy diets more affordable. Recommendations include scaling up investments in climate-smart and nutrition-smart agricultural R&D, new approaches to grain reserves, closer regulation of biofuels policies, trade reform to prevent export restrictions especially, and more adaptive and more nutrition-sensitive social protection programs to protect the incomes and diets of high-risk populations.

This research was funded by the Micronutrient Forum, USA, on behalf of the Standing Together for Nutrition (STfN) consortium through a grant from Global Affairs Canada and the Children's Investment Fund Foundation (CIFF); the Food Prices for Nutrition project funded by the Bill and Melinda Gates Foundation (BMGF) and the United Kingdom government's Foreign, Commonwealth & Development Office. The full study is published as D. Heady and M. Ruel. 2022. "Food Inflation and Child Undernutrition in Low and Middle Income Countries," IFPRI Discussion Paper 2146, International Food Policy Research Institute, Washington, DC.

16 MC12: How to make the WTO relevant in the middle of a food price crisis

Joseph Glauber, David Laborde, Abdullah Mamun, Elsa Olivetti, and Valeria Piñeiro

Originally published June 11, 2022.

The World Trade Organization's 12th Ministerial Conference (WTO MC12) takes place June 12–15, 2022, in Geneva – two years after the pandemic forced members to postpone the meeting's original schedule. In that time span, the world has changed and the need for multilateral approaches to tackling international trade issues has become even more acute. The Russia-Ukraine war has highlighted a variety of unresolved trade issues as poor countries struggle to meet food security needs amid tightening global supplies and high prices, exacerbated by a conflict whose two participants account for 12% of world agricultural exports (on a calorie basis) and threatening 20% of global fertilizer trade.

Two issues of particular concern amid the growing food crisis are export restrictions and public stockholding programs. Agricultural export restrictions have been a major topic in global trade negotiations since the food price crisis of 2007/08, when many countries imposed such measures, particularly on food grains like wheat and rice. Now it is happening again: Since Russia's invasion of Ukraine in February 2022, 23 countries have imposed export restrictions affecting over 16% of global agricultural trade (on a calorie basis).

Long-term WTO negotiations to address export restrictions, including exempting humanitarian aid from such measures, have failed to reach consensus. Public stocks are often used to address humanitarian needs when supply disruptions affect food supplies. Current WTO disciplines allow for public stockholding programs if the support they provide does not distort production or trade – but some members feel that these measures are overly prescriptive and outdated.

Resolving these thorny issues now would be a large step forward, and make progress on longstanding disputes and helping to address current and future price volatility and food insecurity when global supplies are tight and high market prices threaten poor households around the world.

Addressing export restrictions

When global supplies are scarce, countries sometimes limit agricultural exports to keep more food products within their borders and effectively insulate domestic price dynamics from world or regional markets. Such measures include export taxes, licensing procedures, export quotas, controlled releases of sanitary and phytosanitary certificates, various "extended" inspection measures, and complete export bans. But they all share something in common: They are beggar-thy-neighbor policies that increase the cost of imports for trade partners and lead to major coordination issues across countries. They are precisely the type of policies that a well-regulated trading system should discipline to guarantee food-importing countries and their consumers that they will not bear the brunt of global supply shocks.

Unfortunately, under the WTO and its predecessor, the General Agreement on Tariffs and Trade (GATT), policymakers have typically been guided by the nostrum that "exports are good, and imports are bad," and most disciplines are aimed at opening import markets. While WTO rules technically prohibit export restrictions, a broad exception allows them on the condition that they are "temporarily applied to prevent or relieve critical shortages of foodstuffs or other products essential to the exporting contracting party." The WTO Agreement on Agriculture (AoA) itself – formulated during the 1980s period dominated by declining prices and subsidy wars between the United States and Europe – did not address export restrictions.

Yet, when applied to food products, such policies can have devastating impacts on people's livelihoods and health, and on countries' political stability. Export restrictions are mainly used by developing countries, with dramatic consequences for other developing countries (again, the issue of South-South trade relations was not central during the GATT Uruguay Round).

Fortunately, the world has changed over the past 30 years. Low- and middle-income countries have become major food exporters and importers. Unfortunately, though, recent crises (food price crises of 2007/08 and 2011/12, the COVID-19 pandemic, and the current crisis driven by the Russian invasion of Ukraine) have all resulted in spikes in export restrictions.

The poorest countries suffer the most from these policies. Since export restrictions tend to target commodities and staple food products (wheat, rice, soybean, palm oil), their impacts are larger for the least developed countries (LDCs) that depend on such products for their basic needs, as shown in Figure 1: In 2008 and today, 25% to 30% of LDC agricultural imports (measured in calories) have been impacted by such restrictions, while the share is below 15% for developed economies. In the past, such policies have also blocked humanitarian shipments. Indirectly, meanwhile, these measures increase the cost of procuring food on world markets and will continue to hurt poor households and affect programs that address short-term food insecurity.

Figure 1: Share of imported calories impacted by export restrictions (%), by economic group

Note: Numbers include products used for food, feed or energy. Intra-EU trade excluded from computations.

Source: Export restriction tracker

While export restrictions have other flaws and negative consequences, their impact on LDCs alone should place the issue on top of the WTO negotiation agenda – if member states want to be serious about global food security.

In this context, what can WTO members do? Here we outline five steps to address the issue. Two should be undertaken immediately:

1. ***Exempt humanitarian shipments from any quantitative export restrictions or export taxes.*** This issue has been under discussion for years. The policy fix was well-articulated in the 2011 inter-organizations report on agricultural price volatility, and even "decided" by the G20 heads of state that same year: "We decide

that food purchased for noncommercial humanitarian purposes by the World Food Program will not be subject to export restrictions or extraordinary taxes." Yet a decade after, some G20 countries that agreed on this statement in 2011 are still blocking the integration of such language in the WTO framework. Without a clear decision during MC12, there is no point in continuing discussions on the issue.

2. ***Improve the consultation and notification process.*** Ten years ago, it was also clear that WTO members should provide timely and transparent notification when they imposed restrictive measures to provide a basis for common trust and avoid market panics. Yet in practice, few countries provide any such notifications – before or even in the month following action. There is no obvious link between country size or level of income and the notification process. This shows that it is a not a lack of capacity but a lack of political will that remains the main problem. While the IFPRI Export Restrictions Tracker helps in providing key information, it cannot replace the core responsibilities of countries to be transparent about their policies, nor be a substitute for a proper WTO process.

Following those two steps, medium-term measures should include:

3. ***Exempt LDCs and insular economies from agricultural export bans and restrictions.*** With a proper institutional framework and monitoring, the risk of re-exports could be limited, avoiding leakages in the system.

4. ***"Develop an operational definition of a critical food shortage situation that might justify consideration of an export restricting measure."*** Using the same language proposed in 2011, we should operationalize and restrict the usage of the original Article XI language governing such measures, while also outlining clear definitions of the various export restrictions.

Finally, to build trust in the global trading system as a robust pillar of worldwide and national food security agendas:

5. ***Rebalance disciplines between importers and exporters*** by removing all types of quantitative export restrictions, except for exceptional situations, and provide a system of bound commitments on export taxes.

Providing a long-term solution to public stockholding program issues

MC12 also provides an opportunity to find a long-term solution to WTO rules regarding public stockholding (PSH) programs.

Unlike some other thorny issues in the negotiations (for example, market access, domestic support), there is generally broad consensus among members that PSH programs are established to provide food security and should not be used to undermine domestic support disciplines. The question has been whether provisions under the WTO AoA are outdated or inadequate to address food security concerns.

Public stockholding programs are used by some governments to purchase, stockpile, and distribute food to people in need. The WTO AoA explicitly recognizes the need to take account of food security and exempts PSH programs subject to specific criteria laid out in the agreement's Annex 2.

However, some stockholding programs are considered to be trade distorting when they involve purchases from farmers at prices fixed by the governments, known as "supported" or "administered" prices. Also, current public stocks are highly concentrated among a few countries (Figure 2), and a delicate balance must be found between the flexibility required by most LMICs that have limited fiscal capacity to distort global markets (for example, most African economies) and the risk of strong disruptions that could occur when such flexibilities are granted, and used, by large emerging economies.

Figure 2: Grain stocks (maize millet, soghum, rice, wheat) held by selected countries, metric tons

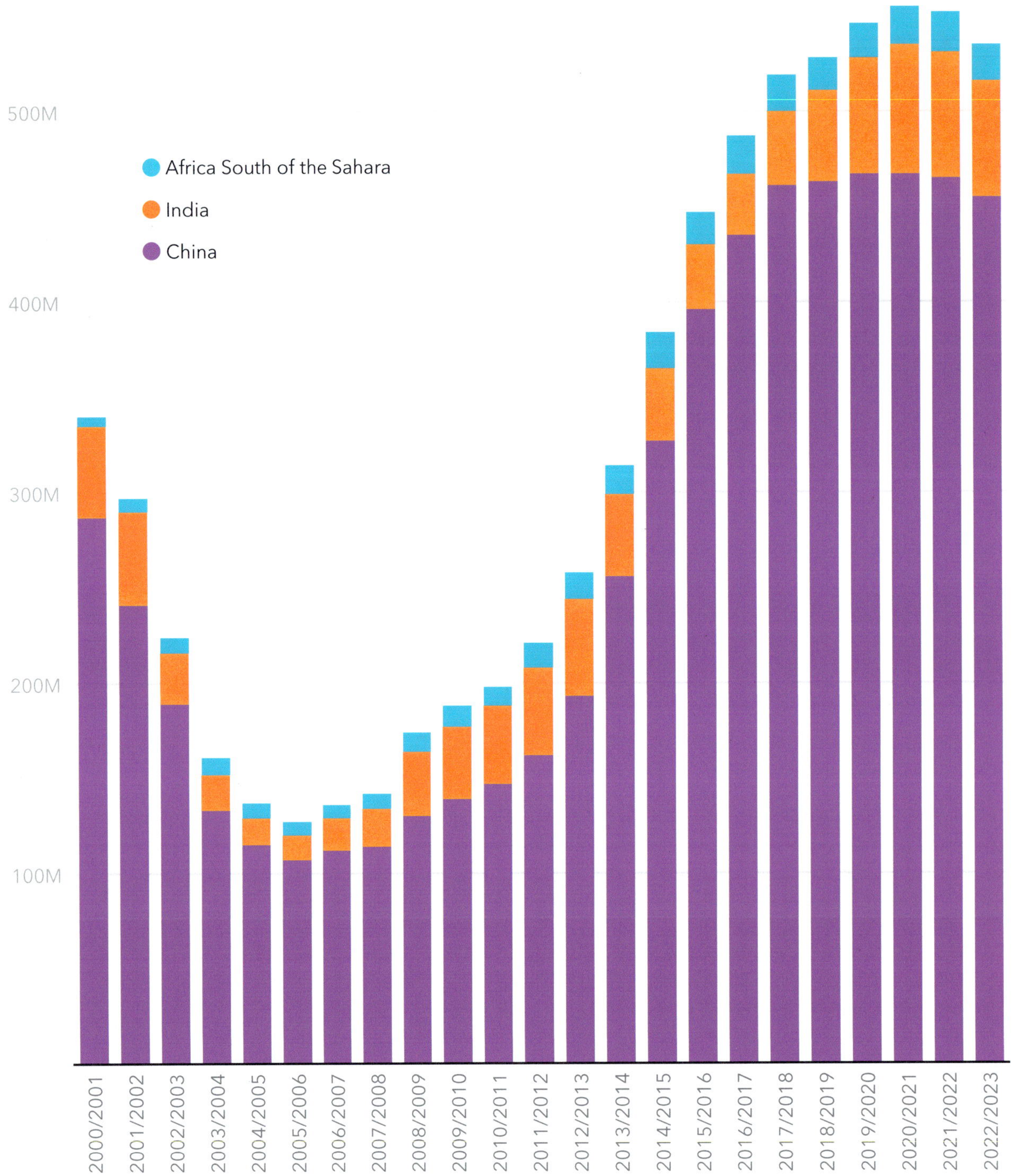

Source: USDA PSD Online

The WTO's interim 2013 Bali decision to exempt PSH programs from legal challenge under certain conditions appears to be the most promising avenue for resolution. More technical fixes, such as updating the methodology for establishing a fixed external reference price, would have broader implications for calculating support under more general price support programs. Moreover, changing the way in which the fixed external price is calculated could raise questions on whether domestic support bindings would need to be adjusted to reflect the new methodology.

One potential avenue would be to expand the Bali decision to include a broader group of eligible foodstuffs than traditional staple crops, but only if the reporting requirements were maintained. Members could consider whether to agree to refrain from challenges under the dispute settlement mechanism support provided by LDCs.

Lastly, transparency is a critical feature underpinning the AoA, and PSH programs should be no exception. Members should be strongly encouraged to report such programs as part of their domestic support reporting requirements. For members who consider such notifications a burden because of lack of timely data, the WTO and other organizations should provide assistance to enhance existing technical capacity.

Conclusions

The global food crisis will take center stage at MC12, and WTO members have an opportunity to resolve long-deadlocked issues in the agricultural negotiations. Progress on other issues will be important as well (for example, domestic support, fishery subsidies, resolving the Appellate Body crisis), but disciplining export restrictions and providing a long-term solution to rules governing public stockholding programs could help protect people around the world from the immediate threat of rising food insecurity and help members better manage food supplies in future crises.

Update

The Twelfth WTO Ministerial Conference was held June 12–15, 2022, in Geneva. Members acknowledged the importance of trade in helping address food security and committed to taking concrete steps to facilitate trade and improve the functioning and long-term resilience of global markets for food and agriculture, including cereals, fertilizers, and other agriculture production inputs. However, little progress was made in addressing outstanding trade issues such as market access, domestic support, or public stockholding. One positive outcome of the 12th Ministerial Conference was the agreement of members to refrain from imposing export prohibitions or restrictions on foodstuffs purchased for noncommercial humanitarian purposes by the World Food Programme.

17 Can the G7 be a force for good in the current global food security crisis?

David Laborde and Carin Smaller

Originally published June 6, 2022.

The Group of Seven wealthy nations (G7), currently led by the German presidency, has put a welcome focus on the global food insecurity and nutrition crisis unleashed by the war in Ukraine, with the most severe impacts falling on vulnerable populations in low- and middle-income countries (LMICs). Two G7 meetings in May produced four separate communiqués, each of them dozens of pages long (the development ministers' communiqué alone was 23 pages!), and a G7-led Global Alliance for Food Security was announced. The G7 meetings coincided with serious efforts on the same front by the United Nations, which has set up a Global Crisis Response Group and convened a UN Security Council meeting on Food Insecurity and Conflict.

The G7 effort can help to ensure a commensurate response to what is turning out to be the worst global hunger crisis in decades, and in so doing help to elevate the G7 itself, whose relevance as an exclusive group of rich and elite countries has been questioned. To realize this promise, however, G7 commitments must be backed up with action – particularly funding.

In 2015, the G7 committed to "lift 500 million people in developing countries out of hunger and malnutrition by 2030." Figure 1 shows the evolution of official development assistance (ODA) from G7 and non-G7 countries to LMICs in the periods just before and after the 2015 Elmau commitment was made. While emergency food assistance has increased, the overall spending needed to achieve the long-term goal of lifting 500 million people out of hunger and malnutrition has stagnated at an average of US$12 billion per year, with the G7 contribution stagnating at $7.5 billion per year. So even before the current crisis, the G7 had failed to make additional resources available to achieve the Elmau commitment.

Figure 1: **Recent evolution of ODA spending for food security, billion US dollars (constant 2018)**

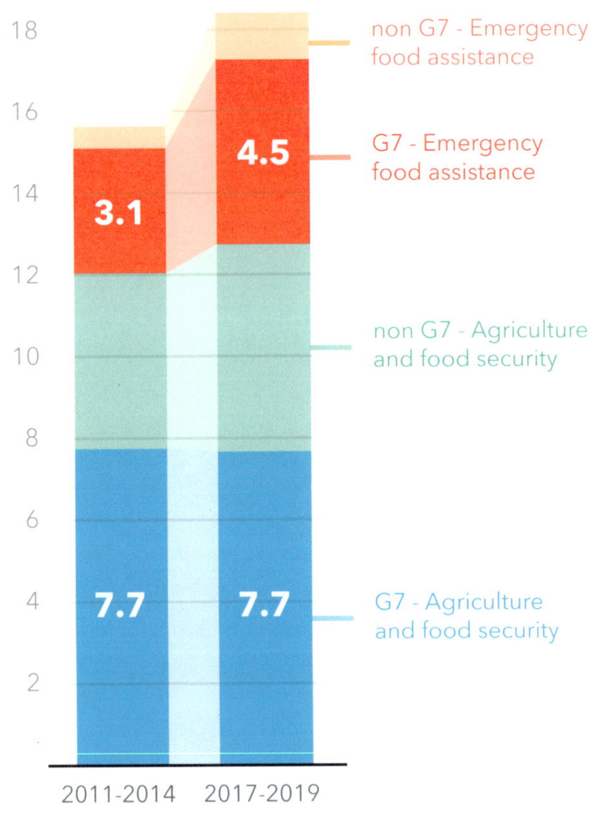

Source: Ceres2030 based on DAC-OCDE

What has changed?

The G7 foreign ministers' May 14 communiqué made global food security, particularly for the most vulnerable, a top foreign policy objective. At the same time, the G7 agriculture ministers sent powerful messages on all aspects of global agriculture and food security, offering concrete financial and technical support to the G20 Agricultural Market Information System (AMIS), which can help create transparency on food, agriculture, and fertilizer markets, thus limiting further disruptions and discouraging individual countries from taking unilateral action, like export bans, that would exacerbate hunger around the world.

The G7 development ministers' communiqué goes a step further, providing more specifics and ambition. The ministers simultaneously committed to a robust immediate response to the crisis alongside a long-term strategy, backed by scientific evidence, to support the achievement of the Sustainable Development Goals by 2030 and build resilience to further shocks. Announcement of the Global Alliance for Food Security, which will work closely with others to respond to the global hunger crisis, particularly through the Global Agriculture and Food Security Program (GAFSP) and the International Fund for Agricultural Development (IFAD), is expected to mobilize $12 billion to respond to the crisis. The communiqué also importantly recognizes other initiatives, including the UN Global Crisis Response Group, the French-led Food and Agriculture Resilience Mission (FARM Initiative), and the Italian-led Mediterranean Ministerial Dialogue on the Food Security Crisis.

The communiqué reaffirms the G7 commitment to give 0.7% of their gross national income (GNI) as ODA, and to reverse the declining trend in ODA to least developed countries. This is particularly important to raise the additional ODA needed, and as a way to put pressure on countries like the United Kingdom, which recently tried to reverse this commitment, and whose aid levels dropped by 51% last year. It also puts G7 countries like the United States and Japan in the spotlight, since they had not yet officially set the 0.7% figure as a national policy target.

The commitment "to strengthen, and not undermine, the long-term resilience and sustainability of agriculture and food systems" reinforces the need to go beyond the crisis and beyond the emergency response. Such a long-term vision in the face of the worst hunger crisis in decades shows an understanding of the interlinkages between a fragile food system and its vulnerability to shocks, not only today but in the future.

The G7's support for the Zero Hunger Coalition, a long-term alliance to eliminate hunger nutritiously and sustainably by 2030, is another important recognition of the need to base action on science, and builds on the outcomes of the 2021 UN Food Systems Summit, including a recognition of the commitments made by companies through the Zero Hunger Private Sector Pledge, which now stands at $458 million in pledges by 43 companies in 48 countries.

Crucially, the rest of the G7 finally follows Canada's lead in supporting a feminist development policy to ensure that all development actions "increasingly target equity and equality of genders and sexual identities." This is an essential policy change to avoid the pervasive neglect of women in agricultural development and to achieve gender equality and women's empowerment.

These stated priorities are ambitious commitments that can make a real difference. Yet the G7 has yet to make sufficient ODA commitments to achieve the longer-term agenda. Without such additional resources, these will remain hollow promises.

How to turn commitments into action?

In their May communiqué, the G7 development ministers recommit to the 2015 Elmau goal, and add that meeting it will be "[b]ased on the sound scientific identification of financing needs and the most efficient interventions and approaches to achieve this goal in a sustainable manner" – a reference to a series of scientific studies by Ceres2030, ZEF, and FAO. These studies found that donors need to double their current ODA contributions to agriculture and food security to achieve the goal (Figure 2).

Specifically, mobilizing an additional $14 billion of ODA for agriculture and food security per year until 2030, and spending that money more effectively in 10 high impact investment areas, is what is needed (see Ceres2030).

The G7 development ministers are to be applauded for the strong engagement and ambition set forth in their communiqué. But the real test for the G7 will be whether these ambitions are matched with the resources required to actually realize them.

At the May meetings, G7 finance ministers pledged an additional $20 billion in 2022 to help Ukraine fight the Russian invasion. This support is essential to help the Ukrainian people. The United States has approved its own $40 billion in assistance to Ukraine. While the aid package is overwhelming in its size and ambition, about 2% of it goes to longer-term agriculture and food security, either at the bilateral level via the Economic Support Fund or through the multidonor GAFSP fund (Figure 3). Unless the G7 also commits to providing resources to those fighting hunger around the world, then the war on hunger can never be won.

Figure 2: Additional public and donor spending needed yearly on average to end hunger sustainably by 2030, billion US dollars

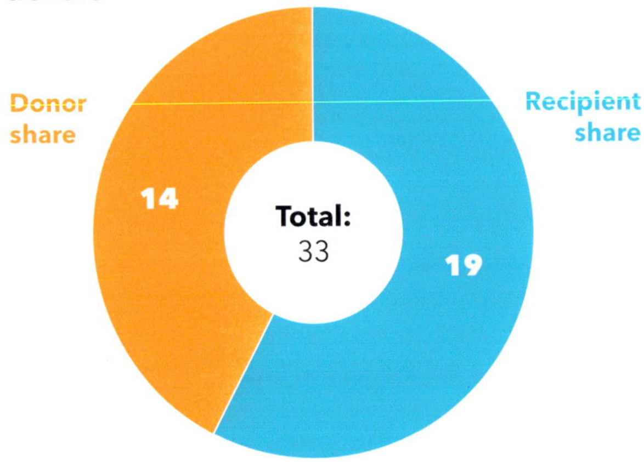

Source: Ceres2030

Figure 3: US response to the Ukraine crisis: Additional Ukraine Supplemental Appropriations Act, May 2022, million US dollars

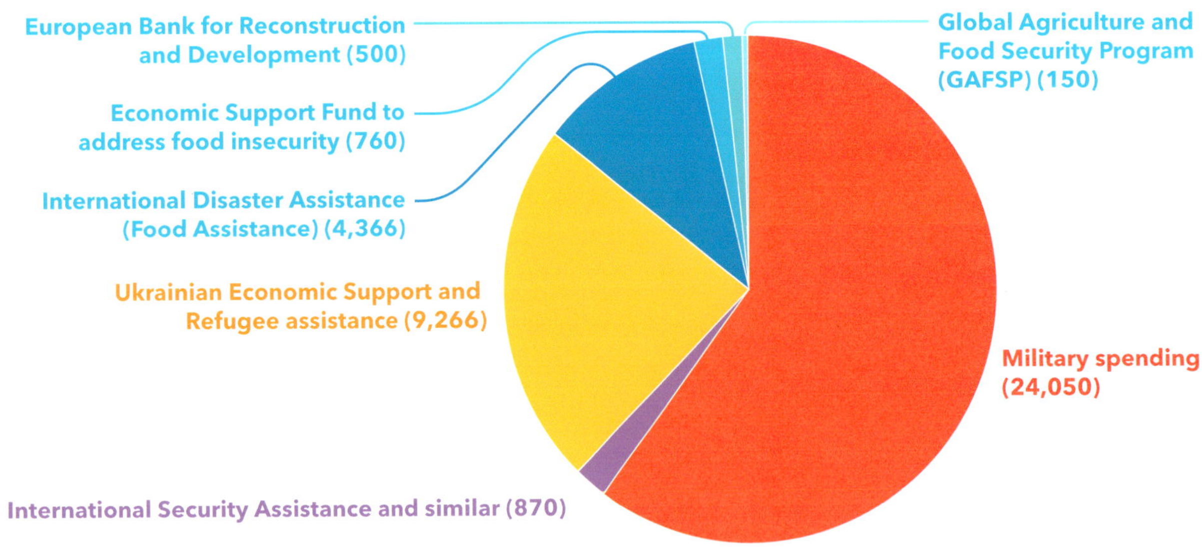

Source: US Congress

SECTION 3

TRADE POLICY RESPONSES AND MITIGATION OPTIONS

18 From bad to worse: How Russia-Ukraine war-related export restrictions exacerbate global food insecurity

Joseph Glauber, David Laborde, and Abdullah Mamun

Originally published April 13, 2022.

Global turmoil and supply shocks can increase a country's vulnerability to food shortages. In the past, countries have often resorted to restrictive trade policies to address food supply disruptions. The Ukraine-Russia crisis is no exception; a number of countries have imposed export restrictions in various forms.

With food prices already high due to COVID-19-related supply chain disruptions and drought-reduced yields in 2021, Russia's invasion came at a bad time for global food markets. Russia and Ukraine alone account for 12% of total calories traded. As the war continues, there is a growing likelihood that food shortages, particularly of grains and vegetable oils, will become acute, leading more countries to turn to restrictions on trade.

Thus this is a dangerous moment. Such measures benefit buyers in domestic markets, but at the expense of net food importing countries. Past experience suggests that these trade measures will put additional pressure on available food stocks, push prices up, and potentially threaten food security for the poor.

In past crises, including the COVID-19 pandemic and the grain price spikes in 2007/08 and 2010/11, many countries restricted the trade of food, including grains and vegetable oils. Worse, export restrictions often had a cascading effect – when one country announced restrictions, others often followed suit, further exacerbating supply problems and creating a panicked atmosphere in global markets as importers sought to secure new suppliers, sending prices even higher.

To help understand the extent and scope of current export restrictions, we use IFPRI's Export Restrictions Tracker to analyze how the war in Ukraine has affected global food trade. This tool has been used extensively to track restrictions put in place during the COVID-19 pandemic. The new version of the tracker includes additional data that tracks the share of restricted exports in terms of kilocalories and US dollars and includes restrictions on exports of chemical fertilizers (though we do not discuss the latter here).

Food export restrictions in response to the war in Ukraine

Since the Russian invasion of Ukraine on February 24, the number of countries imposing export restrictions on food has climbed from 3 to 16 (as of early April 2022) (Figure 1). The total amount of exports affected by the restrictions represents about 17% of total calories traded in the world. Those restrictions include export bans implemented by 16 countries through 29 separate measures, and account for 12.4% of traded calories; and export licensing requirements implemented by 7 countries covering 10 separate measures and accounting for 4.6% of traded calories.

For comparison purposes, we consider the extent of export restrictions in 2008 and 2020. The food price crisis in 2007/08 was driven by a number of factors that pushed agricultural prices to (nominal) record highs, including

Figure 1: Evolution of the share of global trade, in calories, impacted by export restrictions

Note: X-axis shows the week of the year. 1= first week of the year.
Source: IFPRI, Export Restrictions Tracker, accessed April 12, 2022.

back-to-back poor wheat crops in Australia, strong demand for biofuels, high energy prices, a weak dollar, and growing demand in developing economies such as China. By the start of 2008, 8 countries had implemented export restrictions covering approximately 7% of trade. By the end of March of that year, some 15 countries representing 16% of trade had implemented restrictions (Figure 1). Restrictions remained in place throughout most of 2008, with the number of countries implementing them rising to 35 during that summer. A number of commodities were affected, principally wheat and rice.

During the first four weeks of the COVID-19 pandemic, 21 countries implemented export restrictions on a wide range of products. At its peak in May and June of 2020, about 8% of total calories traded were affected. By late summer, most countries had relaxed export restrictions.

Who is imposing export restrictions?

Countries that have implemented export restrictions (Figure 2) represent large shares of some of the key commodities traded in the world. Starting with the epicenter of the conflict, Ukraine has restricted exports to ensure adequate supplies for its population during the conflict; Russia's export restrictions on wheat predate the war and include a floating export tax and export quota. The two countries together account for around 35% of the dollar value of total global trade in restricted items. Measured in terms of caloric value, an important indicator of nutrition and food security, the percentage share is even higher – around 42% of total calories in restricted products. Other notable suppliers imposing export restrictions include Indonesia (ban on palm oil exports), Argentina (ban on beef exports), and Türkiye, Kyrgyzstan, and Kazakhstan (bans on a variety of grain products).

Figure 2: Countries implementing food export restrictions, share of exports affected

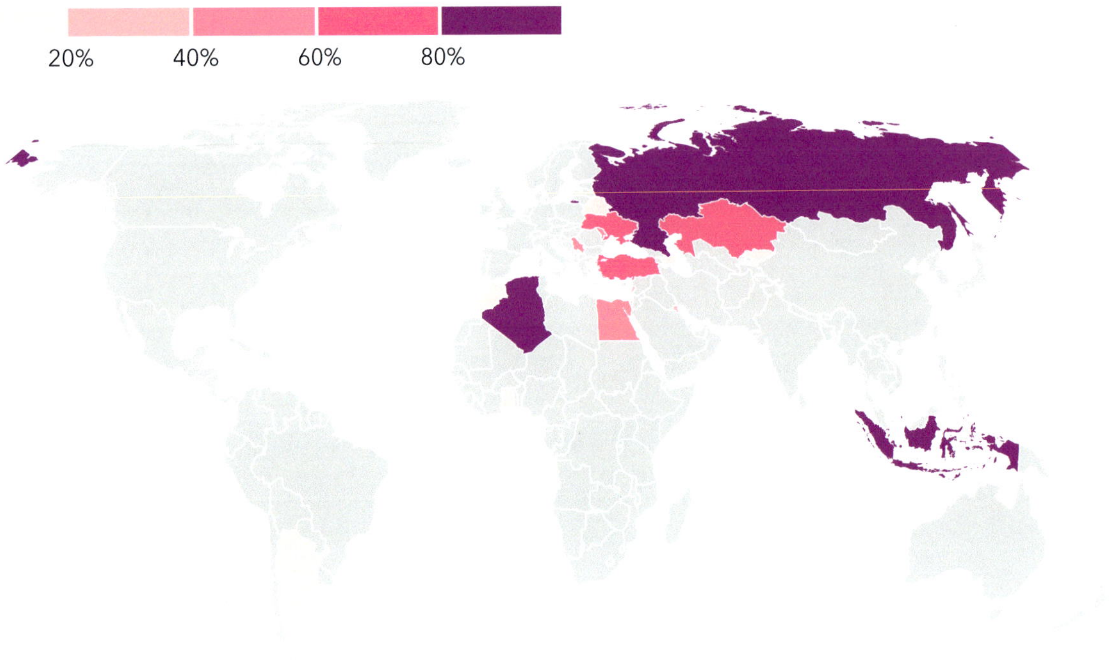

Source: IFPRI, Export Restrictions Tracker, accessed April 12, 2022.

As food price inflation grows around the world, and with no quick resolution to the Ukraine war in sight, past experience suggests that more countries will impose restrictions. This could further raise prices and increase price volatility. Previous research on the impacts of export restrictions during the 2007/08 food price crisis suggests that such policies contributed to 40% of the increase in agricultural prices over the period.

Which products are impacted?

Five agricultural products account for almost 90% of imported calories currently affected by export restrictions: wheat (31% of total calories affected), palm oil (28.5%), corn (12.2%), sunflower oil (10.6%), and soybean oil (5.6%). In terms of total trade in individual products, export restrictions affect 35.9% of wheat exports, 55% of palm oil exports, 17.2% of corn exports, 78.2% of sunflower oil exports, and 5.8% of soybean oil exports. The remaining 10% of traded calories under export restrictions include a diverse basket of commodities such as tomatoes and other vegetables, beef, and poultry.

Which countries are the most affected?

With grains and vegetable oils dominating the list of affected products, it is not surprising that larger importers of wheat, corn, and vegetable oils are most affected. Figure 3 shows the share of imports affected by export restrictions as a share of total agricultural imports of each country (calculated on a caloric basis). Unsurprisingly, import-dependent countries that import wheat and corn from Russia and Ukraine show a relatively high share of restricted calories in their import mix (including Central Asian countries like Mongolia, and North African countries such as Egypt and Sudan). Likewise, export restrictions on vegetable oils, including palm and sun-

Figure 3: Impact of restrictions on importers, share of calories affected, cumulative for early 2022

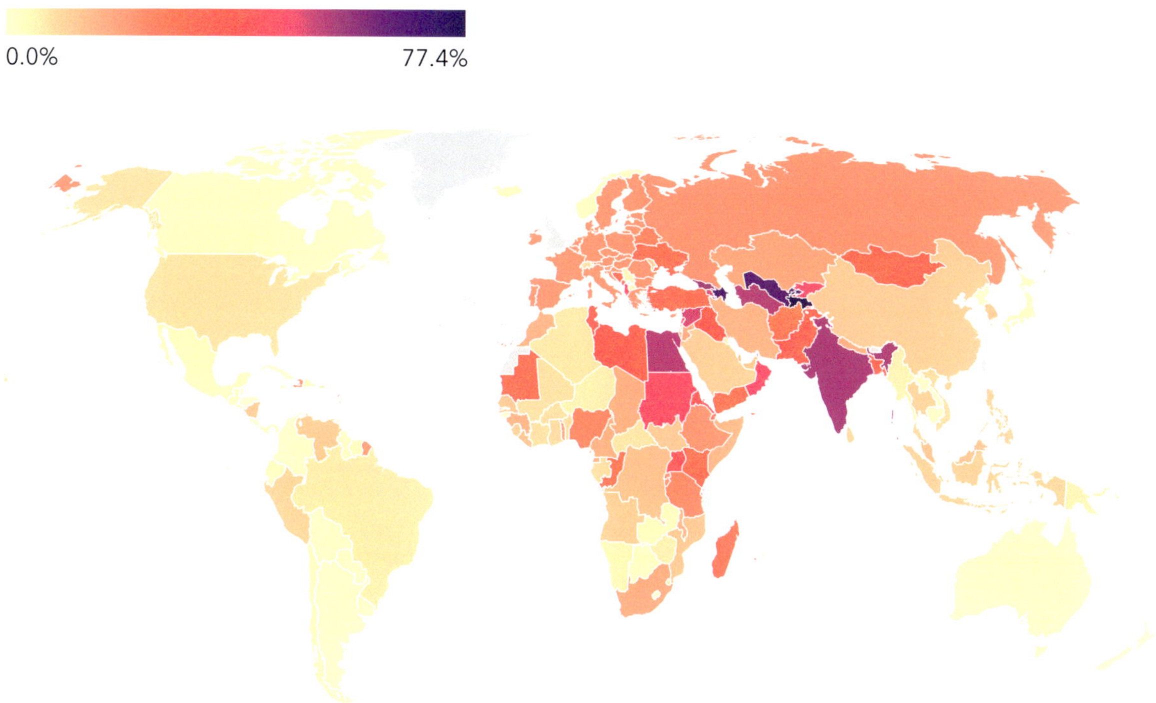

Source: IFPRI, Export Restrictions Tracker, accessed April 12, 2022.

flower oils, have important impacts on countries such as India; Pakistan and Bangladesh are affected both by grain and vegetable oil restrictions.

It is important to recall that many export restrictions are not absolute bans, but rather taxes or other transaction costs that raise the prices of commodity imports but do not prevent them per se. These kinds of restrictions increase consumer prices, but are preferable to outright bans that effectively remove those supplies from the global market.

Conclusions

The principal risk of commodity export restrictions is higher global prices that make it even more difficult for net food importing countries to purchase food. Moreover, export bans tend to be contagious, as other exporting countries follow suit and implement their own bans. The way the global community handled this issue in early 2020, at least for food products, should be an example to follow – most countries refrained from implementing bans, and where implemented, such restrictions were relatively short-lived.

However, the war in Ukraine is taking place in a very different landscape. Initial stocks are low, except for rice, international policy coordination is wavering due to geopolitical tensions, and governments and households are financially exhausted after two years of pandemic – making first-best policies, like the use of social safety nets to compensate for domestic food price increases, more difficult to promote and implement.

Currently, the World Trade Organization (WTO) lacks rules to effectively discipline the use of export restrictions. In the past, WTO members have tried, but failed, to get agreement on disciplining export restrictions on agricultural products, even when trying to create exemptions for basic needs such as humanitarian shipments. The upcoming June WTO Ministerial Conference (MC12) provides an opportunity to address this shortcoming.

In the meantime, countries should cooperate to avoid a wave of detrimental policies like that seen in the 2007/08 food price crisis. Food trade policy can move quickly, but it can have lasting impacts on global poverty and hunger. The Food and Fertilizer Export Restrictions Tracker project provides publicly available data on the global food trade policy environment in the context of the ongoing Ukraine-Russia crisis to aid in curbing this potential problem.

19 Food export restrictions have eased as the Russia-Ukraine war continues, but concerns remain for key commodities

Joseph Glauber, David Laborde, and Abdullah Mamun

Originally published January 23, 2023.

In the weeks following Russia's invasion of Ukraine in late February 2022, several countries imposed export restrictions – including licensing requirements, taxes, and some outright bans – on a variety of feed and food products. These measures helped to fuel war-related disruptions in global markets and contributed to higher prices and increased price volatility. At the peak of the export restriction trend in late May, almost 17% of global food and feed exports (on a caloric basis) were affected (Figure 1) by measures implemented by 23 countries (numbers similar to those reached during the 2007/08 food price crisis: By the end of April 2008, 19 countries had imposed export restrictions, impacting 15.3% of global trade of calories). After May, many countries partially rolled back the measures: By mid-July, the amount of affected trade had fallen to 7.3%, and largely remained at that level over the rest of 2022, though the mix of affected products changed somewhat over the second half of the year.

In this post, we examine the impacts of these restrictive trade measures on prices, supplies, and other indicators. While the pressures that led to the export restrictions have significantly eased and prices of key food commodities have mostly fallen to pre-war levels, the war continues and markets remain volatile, signaling continuing uncertainty and raising concerns that countries could impose restrictions in the future.

Figure 1: Food and feed export restrictions in 2022, by product

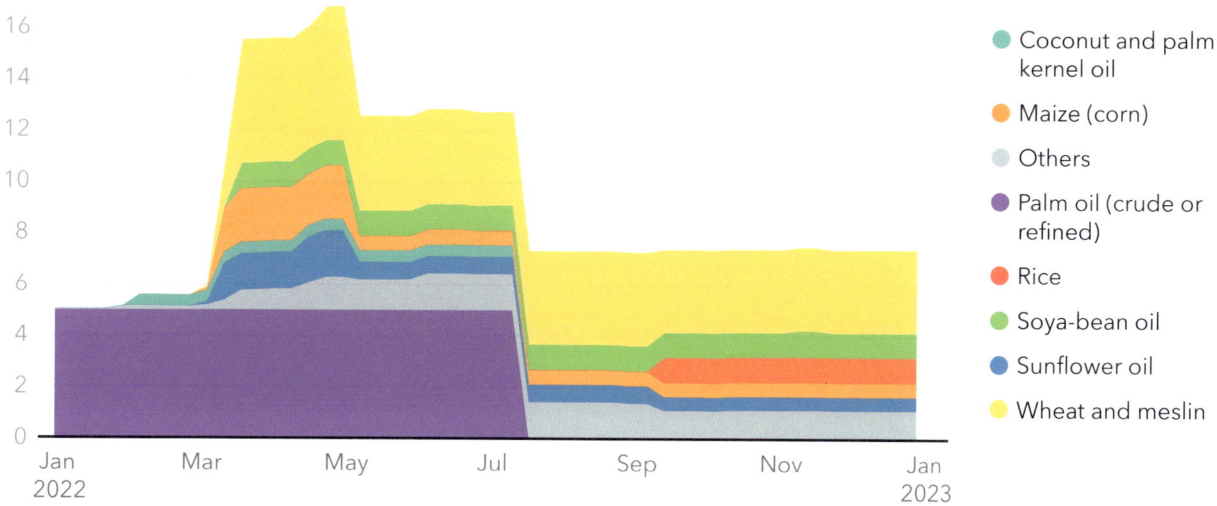

Source: Export Restriction Tracker, IFPRI

Not surprisingly, countries primarily targeted their export restrictions to the commodities most affected by the war – wheat, feed grains, and vegetable oils. Indonesian limits on palm oil exports, for example, accounted for about one-third of all calories affected by such measures, and their May relaxation contributed to much of the subsequent decline in vegetable oil prices. Other grains such as rice have also been affected because of their substitutability. For example, in September 2022, India imposed an export ban on broken rice after exports had risen sharply, driven by high feed grain prices (broken rice is often exported and consumed as a feedstuff).

Export restrictions and the World Trade Organization

Overall during 2022, 32 countries imposed 77 export restrictions in the form of export licensing requirements, export taxes or duties, outright bans, or a combination of measures, according to IFPRI's Food and Fertilizer Export Restrictions Tracker (Figure 2).

The World Trade Organization (WTO) requires member countries to notify it of such measures on a timely basis – but compliance was weak. Of the 32 countries, 26 are WTO members, yet of their combined total of 64 measures, the WTO was notified of only 13 (19%). Of those, 12 notifications were made only after the measure had gone into effect; the average length of time between implementation and notification was more than two months (70 days), and the longest interval was more than eight months (245 days).

WTO members were unable to achieve significant progress on disciplining the use of export restrictions at the 12th Ministerial Conference in June 2022 in Geneva. But members did agree to exempt foodstuffs purchased for noncommercial humanitarian purposes by the World Food Programme from such measures.

Figure 2: Export restrictions and WTO notifications (as of January 9, 2023)

Indicators	Number	Percent
Total number of countries that imposed restriction	32	
Number of non-member countries in WTO	6	
Total number of measures	77	
Number of measures notified to WTO	13	17
Number of measures not notified to WTO	64	83
Number of measures coming from non-member countries	8	
Number of measures notified after start date of restriction	12	86
Number of measures notified in advance or on time	3	21
Average number of days passed between start date of restriction and notification	70	
Max. number of days passed between start date of restriction and notification	245	
Min. number of days passed between start date of restriction and notification	3	

Source: Food and Fertilizer Export Restriction Tracker

Impact of restrictions on exports

The impact of any specific export restriction on market prices depends on its severity, timing, and the level of affected exports relative to overall global supplies. Here we consider two such measures that received much attention when announced in April–May 2022: Indonesia's palm oil restrictions, in force from January to May 2022, and India's export ban on wheat, which took effect May 13 and has continued into 2023. In addition, we examine India's September action banning the export of broken rice and implementing export taxes on indica rice.

Indonesia and palm oil

Palm oil exports account for about 58% of the global trade in major vegetable oils (the others are sunflower oil, soybean oil, and rapeseed oil). On January 27, 2022, in a move to stabilize rising domestic consumer prices, Indonesia announced it was allocating 20% of its palm oil production for domestic use, more than half of which goes toward biodiesel production. The war in Ukraine followed soon after. It had a large impact on sunflower oil exports (sunflower oil accounts for about 13% of the total volume of vegetable oil traded, and Ukraine accounted for about 50% of sunflower oil trade pre-war). As vegetable oil prices rose following Russia's invasion, Indonesia announced that it was banning the export of palm oil and palm kernel oil beginning April 28.

Indonesia accounts for about 60% of global palm oil exports – effectively one-third of the global vegetable oil market. Vegetable oil prices reacted immediately to the export ban, with Chicago soybean oil futures rising almost 4.5% on the news. But the ban was short-lived. It led to protests from producers, including a large number of smallholders, and the government lifted the ban and other restrictions as of May 22, 2022.

The impact of the ban can be seen in Figure 3. Indonesian palm oil exports in May 2022 totaled 193,000 metric tons, 92% below May 2021 levels. Once the ban was lifted, exports recovered somewhat, but cumulative exports were still down 10% through October compared to January-October 2021.

Figure 3: **Indonesia palm oil exports, metric tons**

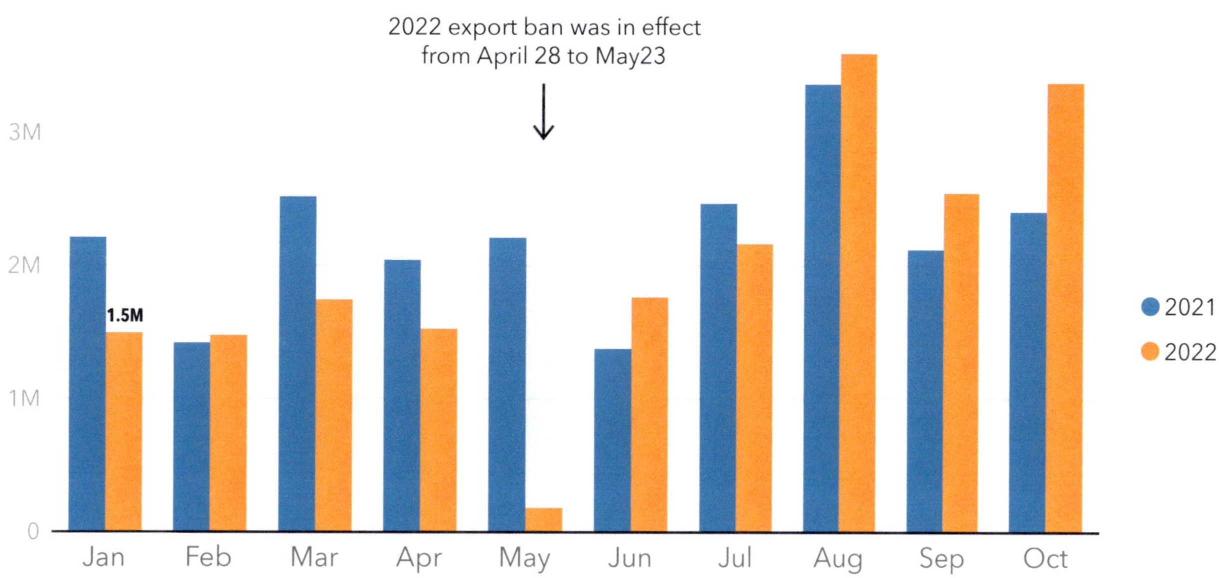

Source: TDM

Figure 4: India wheat exports, metric tons

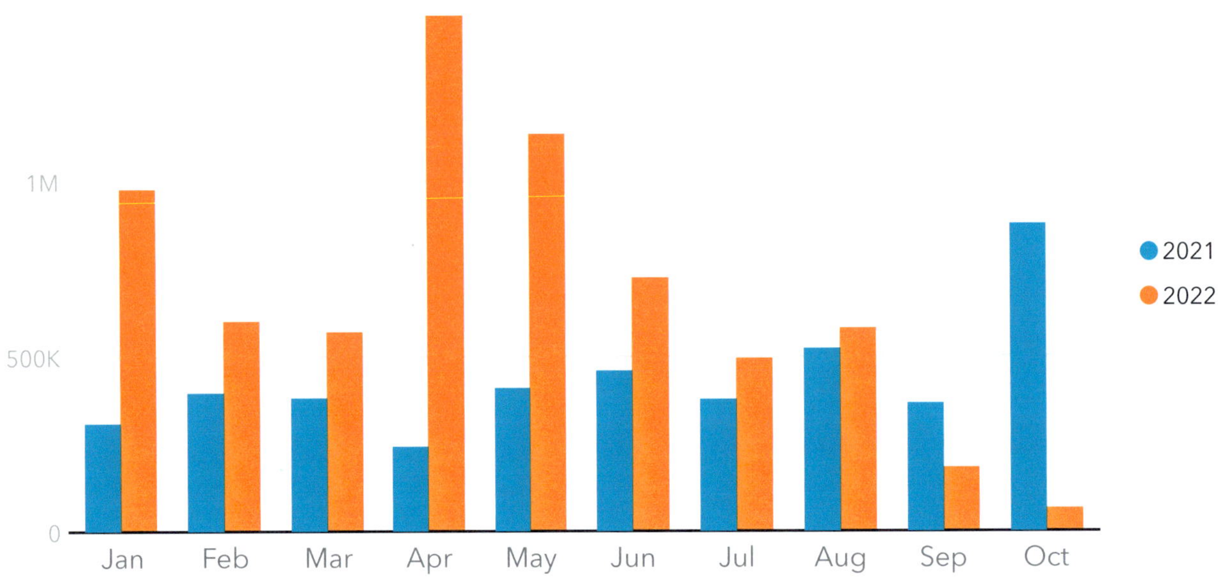

Source: TDM

India and wheat and rice

In 2021, India exported 6.2 million tons of wheat, the largest amount since 2013 and far above the 430,000 tons exported per year, on average, over 2016 to 2020. When the war broke out in February, Indian government officials were confident that the country could export as much as 10 million tons to help offset the resulting global shortage. But dryness in the Punjab region cut wheat yield forecasts and, on May 13, India imposed a ban on wheat exports. Although India is not a major wheat exporter, the move unsettled global markets, with the Chicago benchmark wheat index rising by nearly 6%.

But the export ban turned out to be less draconian than many had feared. Shortly after the announcement, India said that it would honor all export-related letters of credit and that it would "continue to assist neighbours in their hour of need." In fact, as seen in Figure 4, monthly wheat exports continued to exceed corresponding 2021 levels until September, when they fell below 200,000 tons; from May through October, India exported an additional 2 million tons of wheat. Record high prices have caused increased wheat sowings and, with a bumper crop expected in 2023, India is considering lifting its export ban.

In early September, India expanded its trade restrictions, banning exports of broken rice and imposing a 20% tariff on exports of non-basmati rice varieties. India is the world's largest exporter of rice, accounting for over 40% of global exports in recent years. While global rice prices rose throughout 2022, the increases have remained proportionately less than those of other cereals.

Yet these restrictions had only a marginal effect on overall rice trade. While broken rice exports were down by 90% in October compared to year-ago levels, at that point India had already reached a record level of exports for the year. Likewise, the increased tariff on non-basmati varieties had minimal impact on Indian exports (−2%) for the first two months after the tariffs were put in place. Recent reports suggest that India is contemplating removing rice export restrictions.

Impact of restrictions on prices

Figure 5 shows the percent of global food trade affected by export restrictions (on a caloric basis) relative to the UN's Food and Agricultural Organization (FAO) nominal monthly Food Price Index. Though the comparison is largely descriptive rather than analytical, there appears to be a strong correlation between the two series. Prices peaked in May, the same time that over 17% of global trade was hampered by some type of export restriction. Prices then fell from May through July, a period in which some key export restrictions were removed (Indonesian palm oil) or ended up being far less consequential than feared (Indian wheat).

Nonetheless, it would go too far to ascribe causality here – that is, concluding that the removal of export restrictions caused global prices to decline. Several other factors have affected global markets, including good summer harvests and the Black Sea Grain Initiative, which allowed for the partial resumption of exports through Ukrainian ports starting in August 2022. In addition, a drop in global prices can ease domestic concerns and prompt governments to remove export barriers, as we saw throughout the second half of 2022.

Figure 5: Share of global food trade affected by export restrictions vs. FAO Food Price Index

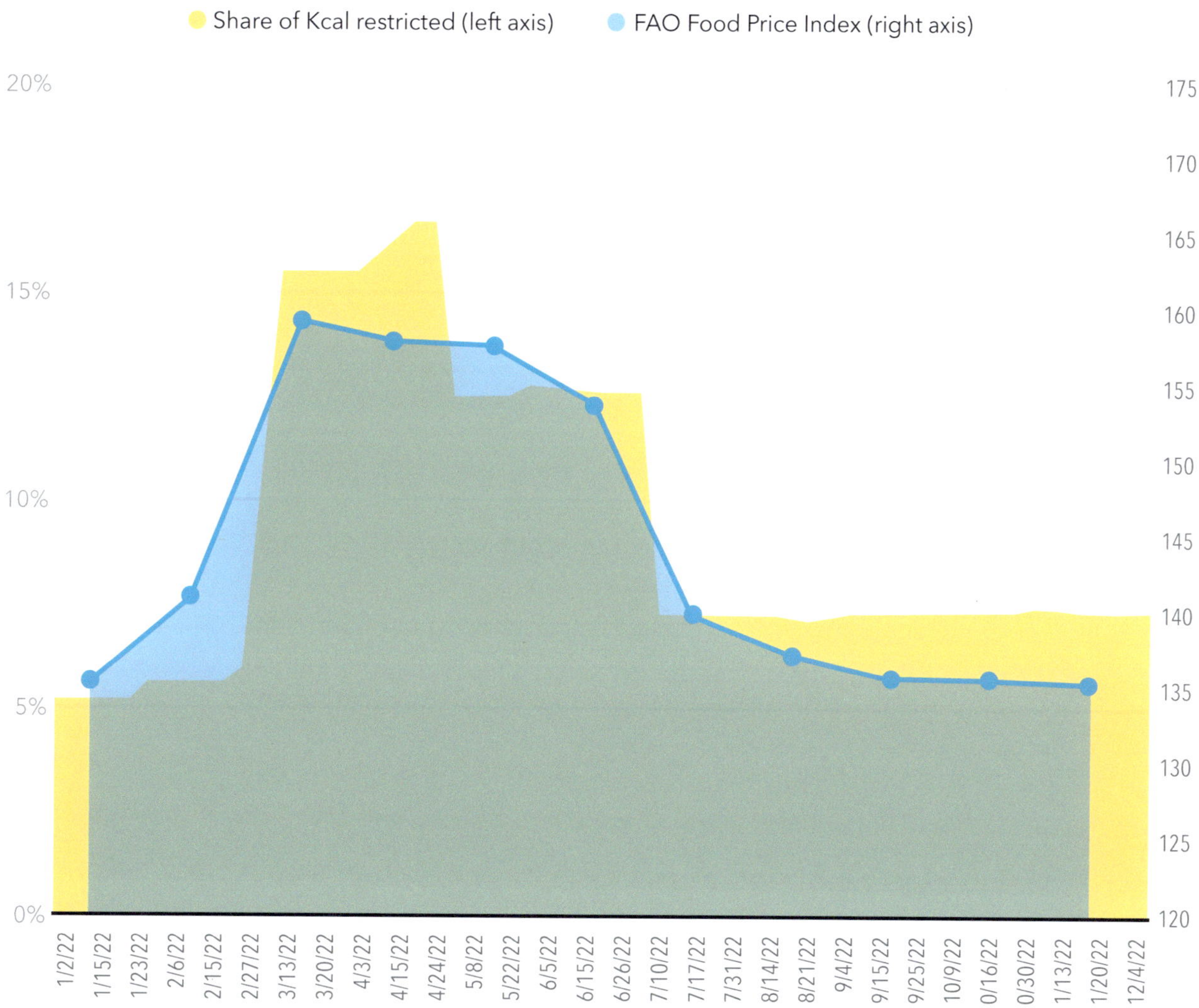

Conclusions

Almost a year after Russia's invasion of Ukraine, global market prices for key food items have returned to pre-war levels. The war continues, but the share of exports affected by export restrictions has fallen by over 50% from its May peak, while the measures themselves appear to be less consequential than many anticipated.

Nevertheless, it is too early to declare the crisis over. Food prices were on the rise before the war began, and they remain elevated compared to mid-2021 levels. Global stock levels remain tight, and with no immediate prospect for peace in Ukraine and drought affecting parts of South America, prices are likely to remain volatile.

Looking forward, improved crop prospects could help build stocks and dampen price volatility. As with wheat and rice, market tensions affecting some crops like maize and soybeans have receded and many export restrictions on them have been lifted. But with global stocks tight, poor crop prospects in 2023 could drive prices up again – and push some countries to consider export restrictions. The lessons of 2022, like 2008 and 2011, show that such policies are ultimately detrimental to domestic producers and global consumers, and should be avoided. When imposed, such restrictions should be applied transparently and the WTO notified in a timely manner.

20 The Russia-Ukraine grain agreement: What is at stake?

Joseph Glauber and David Laborde

Originally published July 27, 2022.

On July 22, 2022, Ukraine and Russia reached an agreement to allow exports of grain and other agricultural products to resume from selected Ukraine Black Sea ports after months of Russian blockade. The agreement comes at a time when storage capacity is reaching its limits, with much of the 2022 wheat harvest and the approximately 20 million metric tons of grains and oilseeds harvested in 2021 remaining in storage – unable to ship because of the blockade.

The deal, brokered by Türkiye and the United Nations, has been widely praised; resuming Ukrainian trade should help ease market prices, consolidating the reductions seen in recent weeks and helping to bring prices back to the pre-COVID-19 levels of 2020. Indeed, wheat prices had already fallen to their pre-invasion level by July 1, while corn prices returned to pre-war levels by July 17 (Figure 1). Still, prices for both cereals remain 50% higher than 30 months ago [July 2020].

Figure 1: Corn and wheat price evolution, US dollars per metric ton

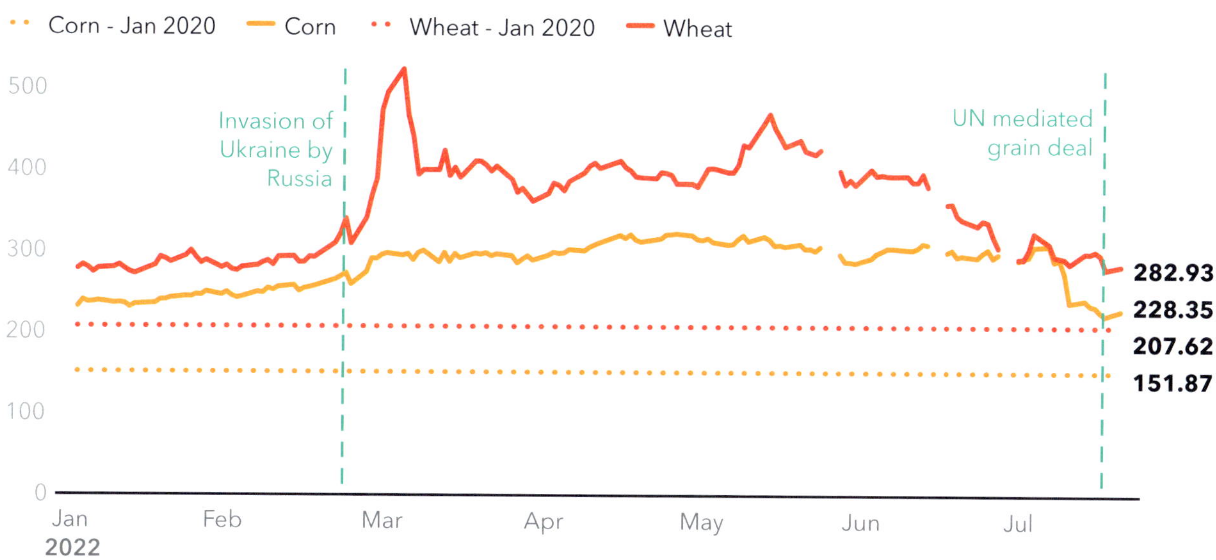

Note: Future prices (+30 days) quoted in Chicago

Source: CBOT

The agreement could help to alleviate rising food insecurity and global hunger concerns. Hundreds of millions of people in the Middle East and North Africa (MENA) region and elsewhere depend on Black Sea wheat imports as an essential part of their diets and are contending with high prices.

However, there is a great deal of uncertainty surrounding the deal. Less than 24 hours after it was signed, Russian rockets struck port facilities in Odesa, damaging grain-loading facilities. Ukraine continues to prepare for the resumption of exports, but the attack has cast doubt on whether insurers will be willing to insure ships operating in the region. Within hours of the attack, futures prices had increased by 5% in reaction. Assuming the deal moves forward amid such risks, what are the likely obstacles and impacts – on global markets, food security in affected countries, and for farmers in Ukraine?

Why is the grain agreement important?

Black Sea ports have been effectively blocked since mid-February, first by Russian military exercises, then by mines placed by Ukraine to prevent possible Russian sea-based attacks after the invasion. Ukraine ships almost 75% of its agricultural exports through the ports on the Black Sea. About half of those exports go out through the ports of Odesa, Chornomorsk (Chernomorsk), and Pivdennyi (Yuzhny) (Figure 2).

Figure 2: Affected Black Sea ports

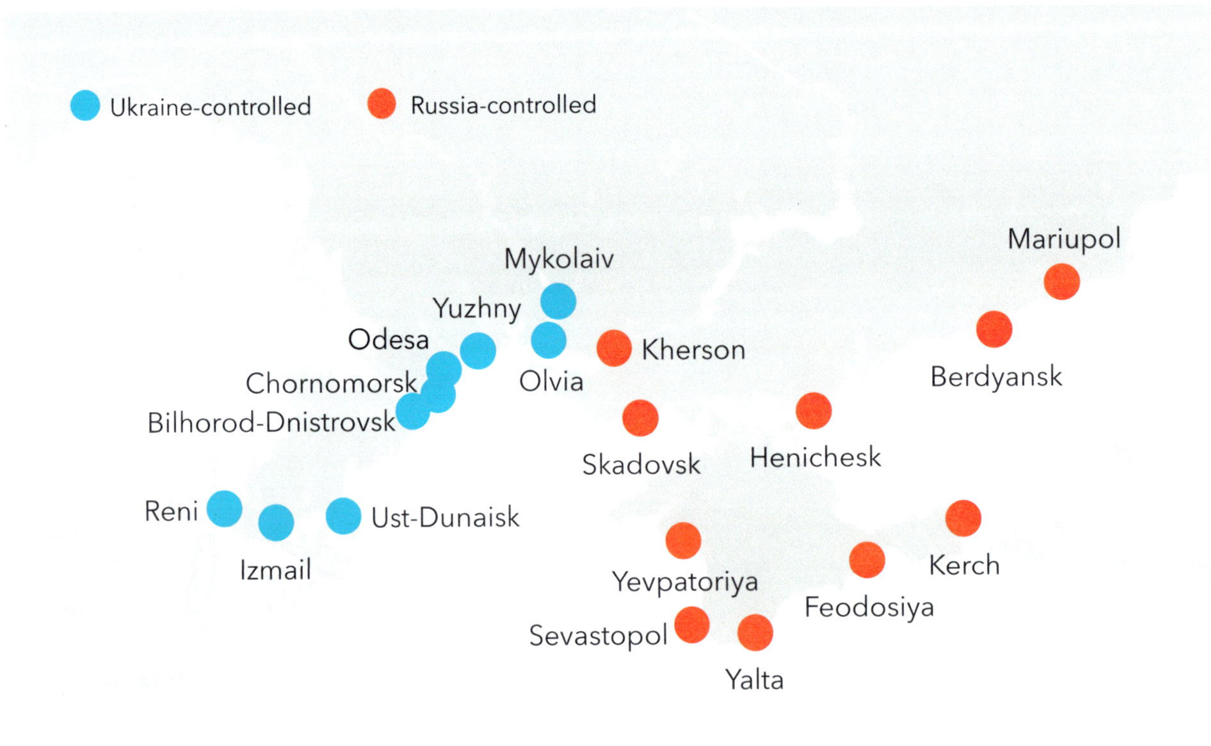

Note: as of July 27, 2022

In 2021, Ukraine exported an average of 4 million metric tons of grains (wheat, maize, and barley) per month and 430,000 tons of sunflower oil. Shipments over the summer and fall can be particularly large, exceeding 6 million tons (Figure 3).

Figure 3: Ukraine's monthly grain exports, metric tons

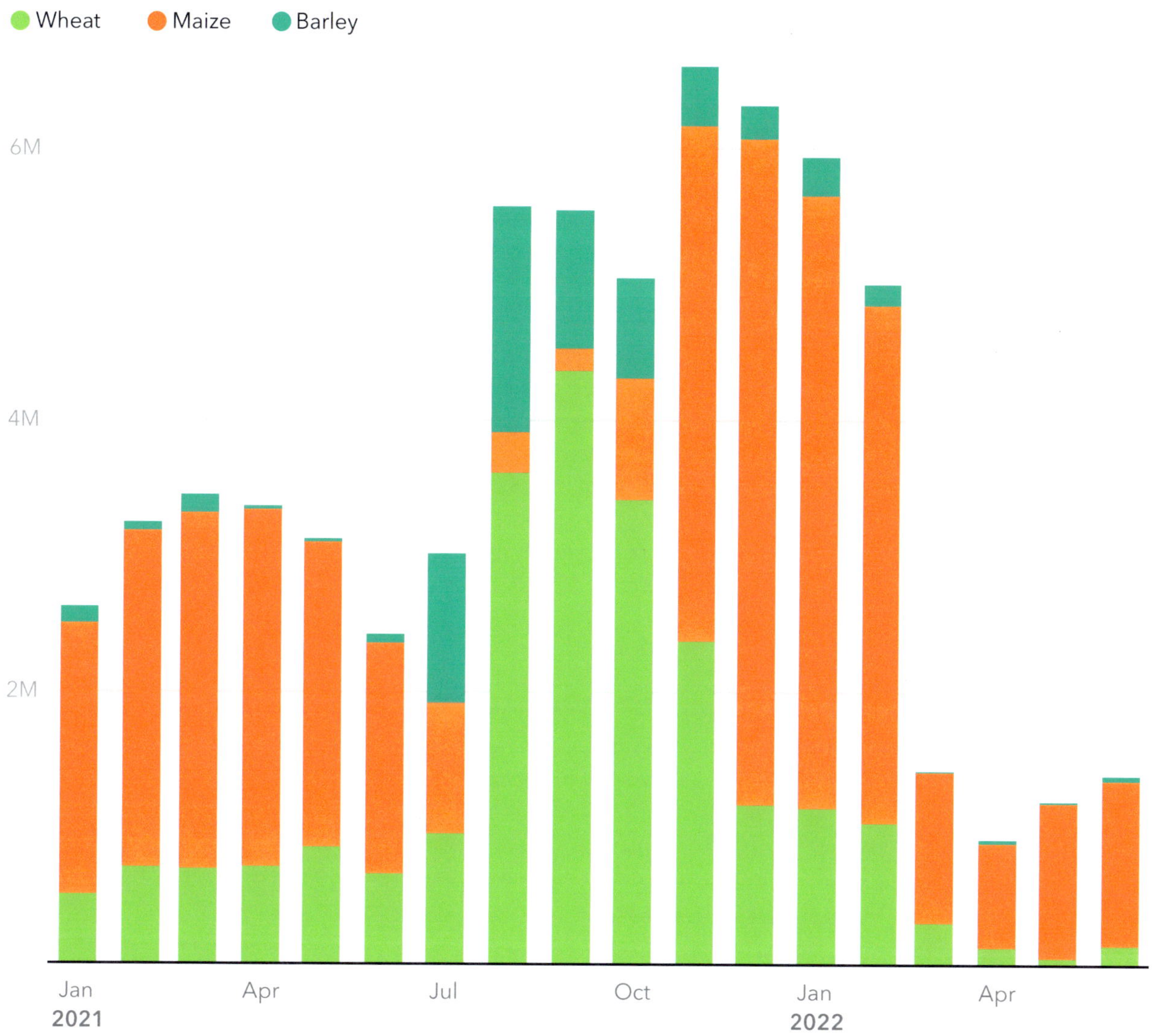

Source: COMTRADE

The blockade resulted in a sharp reduction in exports as grain was diverted to alternate routes by rail and barge to Romanian ports on the Black Sea or through Poland to the Baltic Sea. Despite those efforts, exports have averaged about only 1.5-2 million tons monthly, about one-third the normal level. As a result, about 20 million tons of grain from 2021, including 6 million tons of wheat (13% of 2021 harvest), still has not shipped, causing severe stress on available storage capacity outside of conflict areas, estimated at roughly 24 million tons in early May.

With a 19.5 million ton wheat harvest expected this summer and an additional 38.2 million tons of feed grains expected to be harvested in the fall, analysis suggests storage capacity will be exhausted by late summer (before the harvest of spring-planted crops), forcing the use of temporary storage bags, keeping unharvested crops in the fields, or storing part of the harvest on the ground. Temporary storage could lead to large quality losses in the grain, while the ongoing war threatens further damage to grain transport and storage within the country.

Another issue is the additional costs incurred by exporting from Baltic Sea ports via Poland or through Romanian Black Sea ports. Analysts suggest this may add US$100 per ton or more to the cost of Ukraine's grain. In a competitive global market, those costs must be largely borne by already beleaguered Ukrainian producers. Without access to export markets at sufficient market prices, Ukrainian farmers would have limited incentives to plant in the fall and next spring – thus prolonging the global grain shortage into 2023.

What is in the agreement?

The deal to resume Black Sea exports involves two separate agreements: one between Ukraine and Türkiye covering grain exports from Ukraine, and one between Russia and Türkiye covering food and fertilizer exports from Russia. Under the Ukraine agreement, safe passage of export ships would be allowed from the ports of Odesa, Chornomorsk, and Pivdennyi. Significantly, the agreement does not extend to the port of Mykolaiv, which accounts for about one-fifth of annual Black Sea exports, but is closer to combat zones within Ukraine. Under its agreement with Türkiye, Russia gets assurances that its exports of food and fertilizer will not be affected by export sanctions. This latter point is important, as Russia, like Ukraine, is also a major exporter of agricultural products and fertilizers. The agreement does not extend to natural gas, which is an important feedstock for nitrogenous fertilizers. Finally, the agreement lasts for 120 days but could be extended if agreed to by Russia and Ukraine.

What are the likely effects?

Given that markets remain tense and global grain stocks remain at historically low levels, reconnecting Ukraine to global and regional markets should help – but two questions remain. First, can the agreement be implemented and delivered at scale, allowing exports to return to pre-war levels of 4-5 million tons per month? Reaching such a target will require instituting protocols to guide ships and inspect them, and also convincing commercial operators to reduce shipping rates and currently high insurance premiums. In this context, the July 23 missile attack on Odesa has sent a very negative message.

The primary beneficiaries of the agreement will be Ukrainian farmers. Many are now facing bankruptcy due to very adverse conditions from the ongoing war, the lack of inputs (fuel and fertilizers), and extremely low prices for their crops. Restoring major access to export markets at sufficient market prices (and without the additional costs of shipping overland through Poland or Romania) is essential for their immediate economic survival, and to provide incentives to plant in the fall and next spring – to avoid prolonging the global grain shortage into 2023.

Second, consumers of Ukrainian grains (Figure 4) including food products (wheat) and feed (corn, barley, and also wheat in some markets) also stand to benefit from the deal. These include MENA countries such as Egypt that import to meet the majority of their wheat and feed grain needs – and account for half of the Ukrainian wheat exports – as well as distant countries like China, which rely on Ukrainian corn, among other products, for livestock feed. Among the MENA countries, Türkiye is an important processing center for Black Sea grains, and its food processing industries will also benefit from the normalization of trade. Among the poorest countries and more food insecure, meanwhile, the deal could have two principal effects: Assuming that World Food Program food assistance exports will receive top priority, it could deliver much-needed relief to the 20 hunger hotspots identified by FAO and WFP. And if the agreement leads to lower prices on global markets, it could address the affordability challenge faced by most low-income countries today.

Figure 4: Destination of Ukrainian exports, average 2020-2021, metric tons

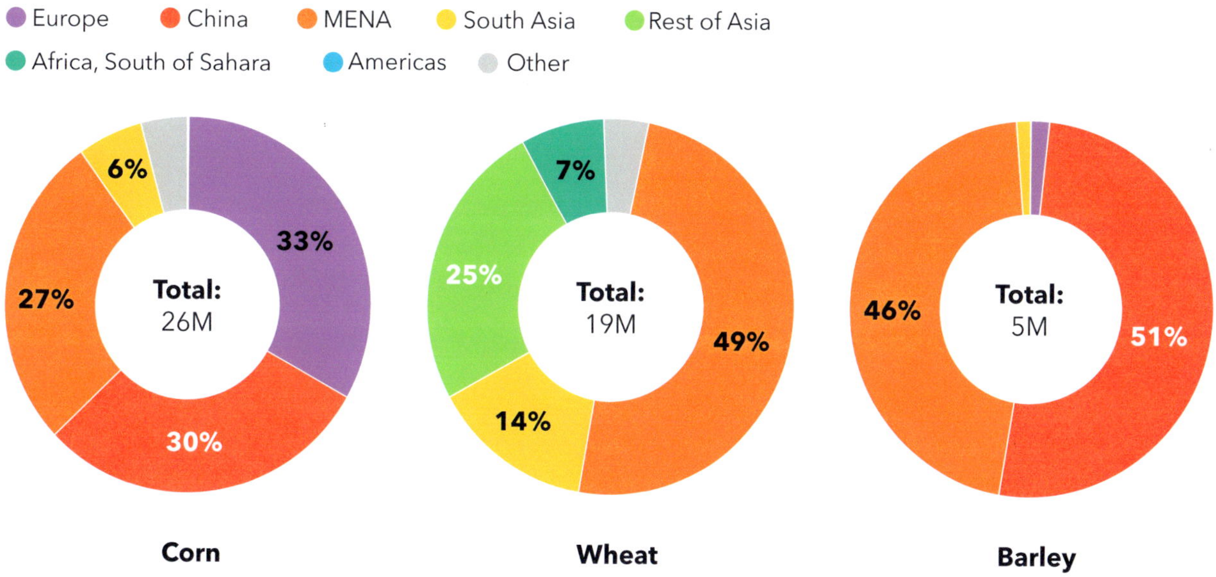

Note: MENA: Middle East and North Africa, includes Türkiye

Source: COMTRADE

Third, Russia will also benefit from safer shipping operations in the Black Sea, a reduction of trade costs that impact its own exports, and potential easing of the indirect effects of sanctions on its exports of food and fertilizer products.

Conclusions

Despite its shaky start, many hope the Russia-Ukraine agreement will result in increased exports of agricultural and fertilizer products. This will help mitigate some of the impacts of the war on the world's poorest households. Is it enough to restore the pre-war status quo? No. The current crisis in supplies of wheat and other agricultural products is driven by many factors and there is a need to rebuild global inventories. This rebuilding process would be greatly accelerated if Ukraine manages to recover its role in the global market. Yet the war is ongoing, with continuing adverse impacts on production and daily uncertainty – a single errant rocket could cause insurers to balk at providing insurance.

In the short run, implementing the agreement at scale and for the rest of year is essential to avoid storage issues that could ultimately drive both prices and production down in Ukraine. The deal is tenuous; nevertheless, it remains the beacon of hope as the first step toward a normalization of trade out of the Black Sea region, and improved food security for hundreds of millions of people.

21 Suspension of the Black Sea Grain Initiative: What has the deal achieved, and what happens now?

David Laborde and Joseph Glauber

Originally published October 31, 2022.

Russia's October 29, 2022, announcement that it was suspending its participation in the Black Sea Grain Initiative – which allows shipments out of Ukrainian ports – was not a surprise; Russia had been skeptical of the deal from the start. But now the supply disruptions will begin again: the move will have a negative impact on Ukraine, its customers, and world market prices – and on global food security, particularly for countries in the Middle East and North Africa (MENA) region. Not only are those countries more dependent on Ukraine as a supplier of wheat and other grains, but they tend to buy more during the winter to supplement their own harvests, which are largely consumed by the end of the year. The renewed interruption in imports could increase food insecurity in these countries and potentially exacerbate political tensions.

The Black Sea Grain Initiative

The Black Sea Grain Initiative, a UN-brokered agreement between Russia and Ukraine, was signed on July 22, 2022. The agreement allowed for exports of grains and related foods to resume from the ports of Odesa (Odessa), Chornomorsk (Chernomorsk), and Pivdennyi (Yuzhny), which had been effectively blocked since mid-February, first by Russian military exercises, then by mines placed by Ukraine to prevent possible Russian sea-based attacks after the invasion. Ukraine ships almost 75% of its agricultural exports through the ports on the Black Sea. About half of those exports go out through the three ports covered by the deal. The agreement has also paved the way for additional measures aiming to ease trade costs in the area by reducing insurance premiums, including through special facilities.

The 120-day agreement was scheduled to expire in mid-November, and while many hoped for its renewal, Russia has been critical of it, and almost since the signing had threatened to end it once the official expiration date arrived. In addition, the grain shipments have faced slowdowns in recent weeks. Clearing times increased from an average of 9 days in September between departure from Ukrainian ports to the finalized inspection in Türkiye (Turkey), to an average of 16 days in October, leading to an increased backlog of ships.

As of October 28, over 9.3 million metric tons of grains, oilseeds, and other foodstuffs had been exported under the agreement. The deal had allowed Ukraine to more than double its exports compared to the pre-deal level – though it is still operating at 50% of its pre-war 2021 level (Figure 1). So, while the deal did not solve all the problems surrounding food exports from the conflict zone, it was important in easing the pressure on regional markets and on Ukrainian farmers unable to move their products. Terminating the agreement now – when Ukraine typically ships a sizable portion of its annual export volumes, particularly wheat – will pose significant problems for Ukraine and its customers.

Figure 1: Ukraine's grain exports, metric tons

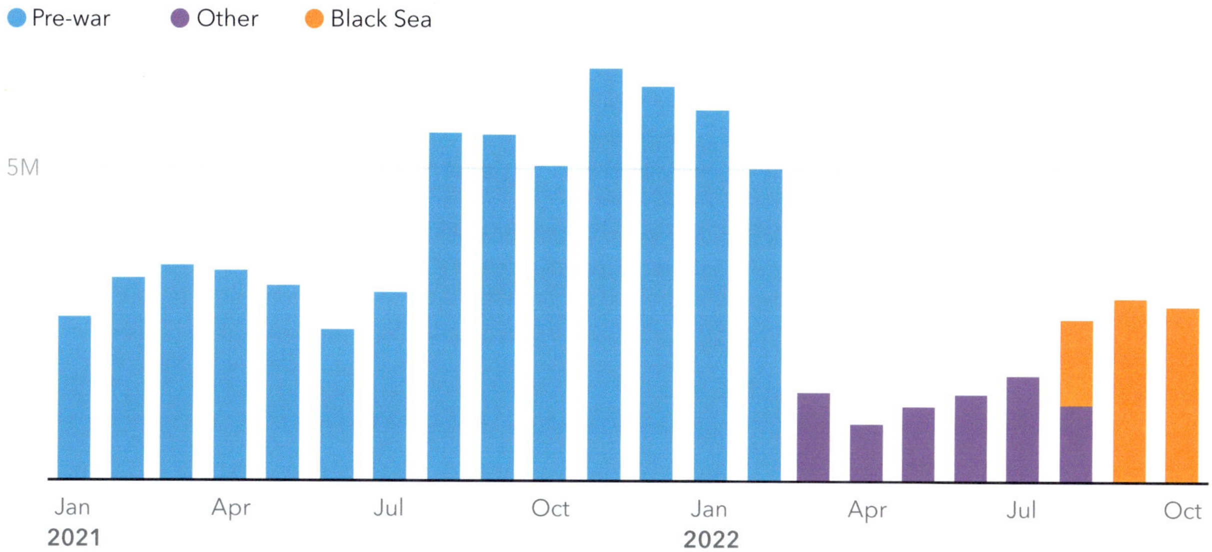

Note: Black Sea = exports reported through the Black Sea Grain Initiative. Grain exports include barley, maize and wheat. Other exports available only through August 2022.

Source: COMTRADE; UN Black Sea Grain Initiative as of 28 October 2022.

Figure 2: Ukraine's wheat exports, comparison of 2021 and 2022

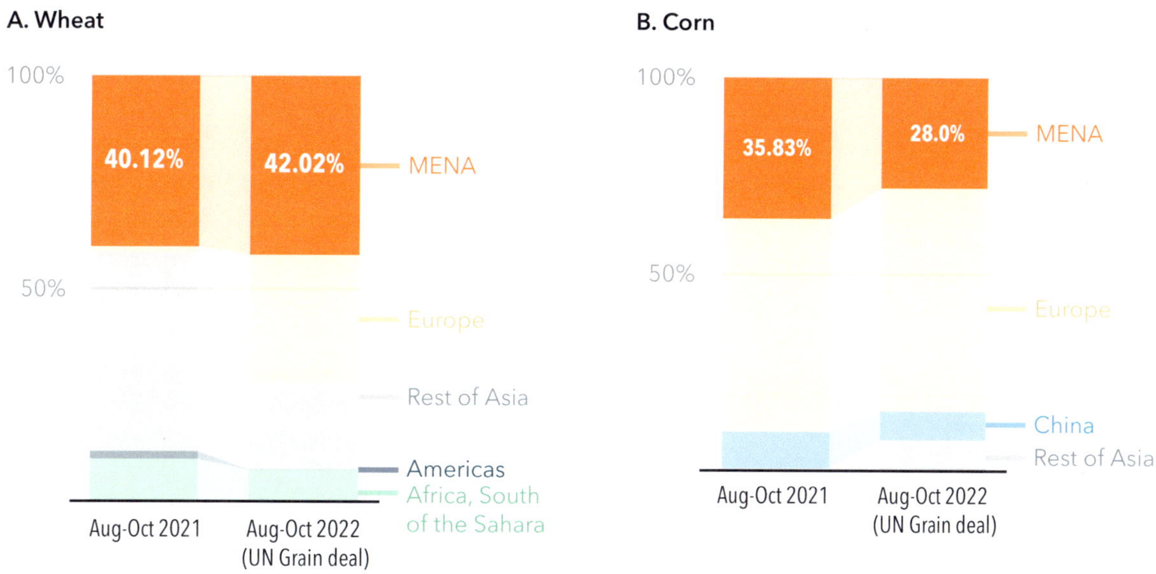

Note: MENA includes North Africa, Western Asia and Türkiye. 2022 numbers only include exports through maritime shipments done unde UN grain deal agreement

Source: COMTRADE, UN

Where are the exports going?

Russia has claimed, among other things, that the deal mainly benefits high-income countries. However, this criticism fails to account for distortions in export patterns caused by the war. Shipping was shut down from February through July, the period when Ukraine typically ships most of its maize, the majority of which goes to Europe. When the deal took effect, maize exports swelled (4 million tons from August to October vs. 1.36 million tons for the same period in 2021). However, both European and MENA countries saw their maize imports increase dramatically during this period and received roughly the same proportions as in 2021, as shown in Figure 2.

Meanwhile, some of the poorest countries, in particular in Africa south of the Sahara, have received the same share as last year in wheat exports. In addition, according to the UN Black Sea Grain Initiative Joint Coordination Centre, about 150,000 tons of wheat has been exported through the World Food Programme to poor countries in the Horn of Africa (Djibouti, Ethiopia, Somalia, and Yemen) and to Afghanistan.

What will the impact be if the agreement is not renewed?

By the time the agreement was signed in late July, market prices for wheat, maize, and other commodities had already fallen significantly from historical highs reached in mid-May. The decline was due to many factors including a strong dollar, declining transportation costs, weakening global demand, a strong corn harvest in Brazil, and less-than-anticipated drought damage to wheat harvests in Western Europe and North America.

Figure 3: **Wheat and corn prices, US dollars per metric ton**

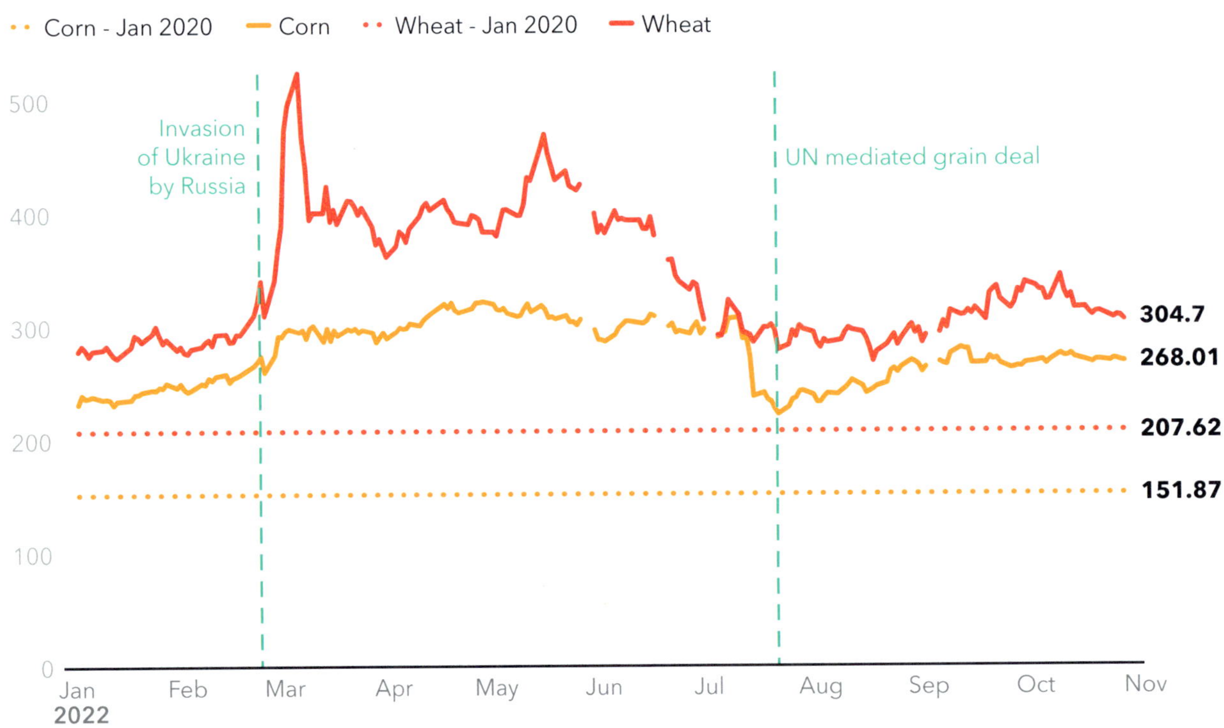

Note: Future prices (+30 days) quoted in Chicago
Source: CBOT

Then, after some uncertainties in September pushed prices up again (Figure 3), hopes that the agreement would be extended lowered pressure on prices during the past few weeks. Before Russia suspended the deal, grain prices had stabilized at pre-war levels; however, they remain 50% higher, or more, than January 2020 levels.

Now, with Ukraine exports again on hold, the deal's suspension will increase pressure on world prices, especially for wheat, whose projected global inventory levels remain at a historically low level. The suspension will also immediately disrupt key grain supplies for MENA countries – in particular Türkiye, Egypt, Lebanon, Sudan, and Yemen, which were benefiting from the resumption of Ukraine exports.

Futures prices for wheat and maize were up 5% and 2%, respectively, following announcement that Russia had terminated the deal. Higher global market prices mean consumers around the world will pay more for imports.

Unfortunately, the suspension also means that Ukraine producers will see little if any benefit from higher prices. Significantly less grain will be moved out of Ukraine, creating more pressure on storage facilities that are already at capacity as Ukraine's farmers harvest spring-planted crops. Lack of storage facilities and restricted export opportunities mean lower prices for farmers.

Lower prices will bring some Ukrainian farmers to the verge of bankruptcy and create further disincentives to plant for next crop year. Even prior to the suspension, Ukraine's Agriculture Minister Mykola Solsky said farmers would sow 20% less winter wheat this fall. A drop in 2023 production would mean the third straight year of disruptions to the Ukraine wheat crop. As Ukraine typically accounted for about 10% of global wheat exports before the war, the effect on global markets is akin to back-to-back droughts over three years in a major wheat-producing region, and it likely means that global stocks will not recover for at least another year. Tight stocks means continued high prices and volatile markets.

Conclusions

Russia's suspension of the Black Sea Grain Initiative is a setback for efforts to reduce the impacts of the war in Ukraine on global consumers, particularly those in the MENA region most reliant on grain imports from Ukraine ports – once again threatening food security in those countries. Efforts are underway by the United Nations and Türkiye to bring parties back to the negotiating table, but for the moment, prospects look dim. The short-run effects will include higher global prices and a continued disruption in trade patterns for those countries that have depended on Ukraine for grain and oilseed imports. The suspension will also hurt Ukraine's producers, meaning that market disruptions will continue to have global impacts into 2023 and perhaps beyond.

Update

Despite threats of termination by Russia in late October 2022, the Black Sea Grain Initiative was renewed in November. The current agreement is scheduled to expire in late March 2022. From August 1, 2022, to January 4, 2023, some 16.6 million metric tons of agricultural products had been shipped from Ukraine ports on the Black Sea, including 7.6 million tons of maize, 4.7 million tons of wheat, and over 700,000 tons of barley. Black Sea exports combined with exports through the "solidarity lanes" (rail, river, and truck routes out through western Ukraine to EU destinations) have allowed Ukraine's exports to recover significantly since the start of the war, though exports still remain below past monthly levels.

22 How sanctions on Russia and Belarus are impacting exports of agricultural products and fertilizer

Joseph Glauber and David Laborde

Originally published November 9, 2022.

The sanctions imposed by the European Union, United States, Canada, and other countries on Russia and Belarus following Russia's February 24 invasion of Ukraine included restrictions on banking, trade, technology transfers, and specific individuals. These came on top of earlier sanctions on both countries – on Russia, in response to its 2014 annexation of Crimea, and on Belarus, in response to human rights violations in 2020 and its forced grounding of Ryanair flight 4798 to seize a dissident journalist in 2021.

The current sanctions regime aims to spare the agriculture sector, including inputs, to avoid side effects on global food security. Russia is a major global supplier of key staples, including wheat, and both countries are important fertilizer producers and exporters. Shutting down trade in those items would have disastrous consequences for global markets and on agriculture and food supplies in countries reliant on them.

But given the complexity and scope of the sanctions on Russia and Belarus, and the reluctance of many countries and companies to do business in either country, are the agriculture exceptions working? In this post, we examine how the sanctions aimed at Russia and Belarus have affected exports of agricultural products and fertilizers. The picture is mixed, with exports of some items affected by the war's outbreak, then recovering, while others have slowed significantly.

Carve-outs for food and fertilizer

Both the United States and European Union have specified "carve-outs" for the food and fertilizer sectors. Only luxury food products, like caviar, are prohibited from being imported into the United States or EU. In July, the US Treasury Department issued guidance to clarify that the production, manufacturing, sale, or transport of agricultural commodities including fertilizers, agricultural equipment, or medicine originating from Russia and for destination to the US or third markets were exempt from sanctions.

Similarly, EU sanctions exclude Russian exports of food supplies and fertilizers to both EU and non-EU markets. While this policy aims to avoid targeting fertilizer overall, potash is an exception. Its critical role for the Belarus economy (10% of all exports) and government led the EU to impose specific sanctions on potash and companies involved in its production, banning imports in the EU and restricting the capacity of Belarus to use EU territory for transit – in particular through Lithuania.

To avoid any diversion effects or re-exports of Belarus potash under a Russian flag, the EU extended the restriction to Russian potash as well. However, this last measure does not limit the trade between Russia and third parties directly. Rather, it restricts access to European ports for shipping companies operating with Russia. In addition, while ships carrying fertilizer and food from Russia should have access to European ports, individual member states also retain a degree of discretion regarding the implementation of such waivers.

But the carve-outs may not work perfectly. Even if sanctions do not directly target food and fertilizer exports, some do target individuals – such as oligarchs – who may have business interests in these sectors, and in general they can make trade more difficult. For example, importers may choose not to purchase from Russia and Belarus because of the added costs of doing business (for example, having to deal with more restrictive banking regulations), or due to overcompliance, that is, the perceived risk, even if erroneous, of being ensnared in financial sanctions for having done business with these countries. Some companies have also stopped trading with Russia due to negative public opinion, or as a result of concerns expressed by their employees.

To avoid such adverse effects, countries including the United States have started to issue "comfort letters" to facilitate exports of some essential products from Russia by clarifying that sanctions are not in place.

Meanwhile, Russia's own actions have created problems with agricultural exports. Since last year, Russia has restricted exports of fertilizer and agricultural products through export taxes, licensing requirements, and actual bans. For example, from March 14 to June 30, 2022, Russia imposed a ban on wheat exports to other countries of the Eurasian Economic Union (EEU) in order to prevent exporters from using the EEU free trade zone to circumvent a significant boost – up to US$130 per metric ton – in Russia's wheat export taxes. In addition, Russia put export licensing requirements in place in 2021 for nitrogenous-based fertilizers (including compound fertilizers). Even if the restrictions fall short of actual bans, they directly limit exports and add costs to Russian imports and thus can adversely affect demand.

Assessing the impacts of sanctions

Assessing if ongoing sanctions on the Russian and Belarusian economies do end up impacting exports of food and fertilizer is complicated by several factors. First is the lack of export data: Both countries stopped reporting trade data to the United Nations and other data partners shortly after the invasion in February. Second, for all the commodities considered, prices have been increasing since 2021, reducing global demand. Thus some decline in Russian and Belarusian exports is to be expected.

An often-used approach to estimate exports in the absence of export data is to analyze mirror trade data from importing countries. For example, by totaling the quantity of imports of Russian potash to its trade partners, one obtains an approximation of total Russian potash exports over a given period (ignoring lags due to the length of transport).

For this post, we used the mirror approach to examine imports from Russia and Belarus for countries that have reported data in 2022, comparing January–August 2022 with the same period in 2021. While the measures are suggestive of relative trade volumes, this approach has limitations. Many importing countries – particularly developing countries in Africa – do not necessarily report trade data on a regular basis, so a complete picture from mirror trade may lag several months or even years behind.

Figure 1 provides an overview of our findings for Russia: Most exports are down compared to 2021, except for phosphate products, which are up in part due to reduced exports from China, which restricted phosphate exports during the first part of 2022. However, the largest reductions occurred in the first four months of the year – affected by the beginning of the war, the large disruptions in Black Sea shipping, and Russia's own export restrictions. Data from the next four months, from April to August, show more limited decreases and even some increases, even though enforcement on most sanctions was starting in earnest during this time.

Figure 1: Imports from Russia by reporting importers, 2022 vs. 2021, percent change

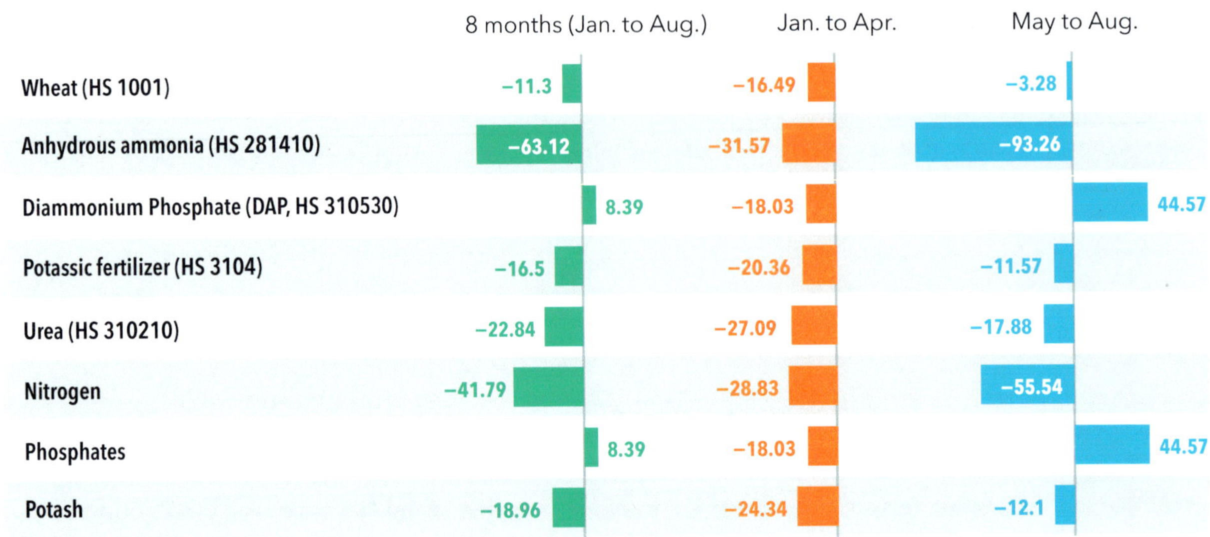

Note: 88 reporting importers.

Source: TDM

Russian wheat

The data show that Russian wheat exports fell by about 11% from 2021 levels between January and August. This decline largely reflects the drop in exports that occurred in the early weeks of the war, when shipping in the Black Sea largely came to a standstill. The data show that trade has recovered somewhat since then (Figure 2).

Figure 2: Cumulative wheat imports from Russia by reporting importers, 2022 vs. 2021, million metric tons

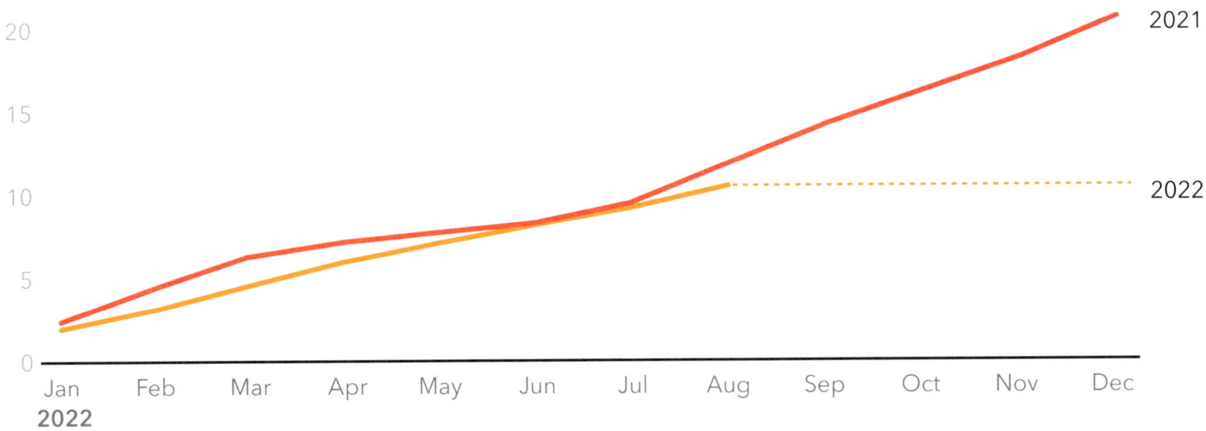

Note: January through August. Data from 34 reporting countries.

Source: COMTRADE

Russian and Belarusian fertilizer

Fertilizer imports from Russia also declined compared to 2021 levels between January and August: Potash and urea imports decreased by 16.5% and 22.8% respectively. As with wheat, most of the decline occurred in the early weeks of the war, and monthly import numbers have recovered somewhat since then. Meanwhile, diammonium phosphate (DAP) imports from Russia are running over 8% above 2021 levels.

The most significant declines have occurred in anhydrous ammonia imports from Russia (down 63%) and potash imports from Belarus (down 50%) (Figure 3). These products are traditionally shipped through ports in the EU (potash from Belarus) or Ukraine (Russian ammonia). Last year, Russia shipped about 4.4 million metric tons of ammonia, about 45% of which went through Baltic ports and about 55% via the world's longest ammonia pipeline from Tolyatti on the Volga to the port of Pivdennyi (Yuzhny) near Odesa, Ukraine, which is currently inactive due to the war.

Ammonia exports pose another problem: The chemical is used both in fertilizers and for non-fertilizer applications, making the implementation of sanctions and associated waivers a potentially tricky prospect.

Figure 3: Cumulative potash imports from Belarus by reporting importers, 2022 vs. 2021, million metric tons

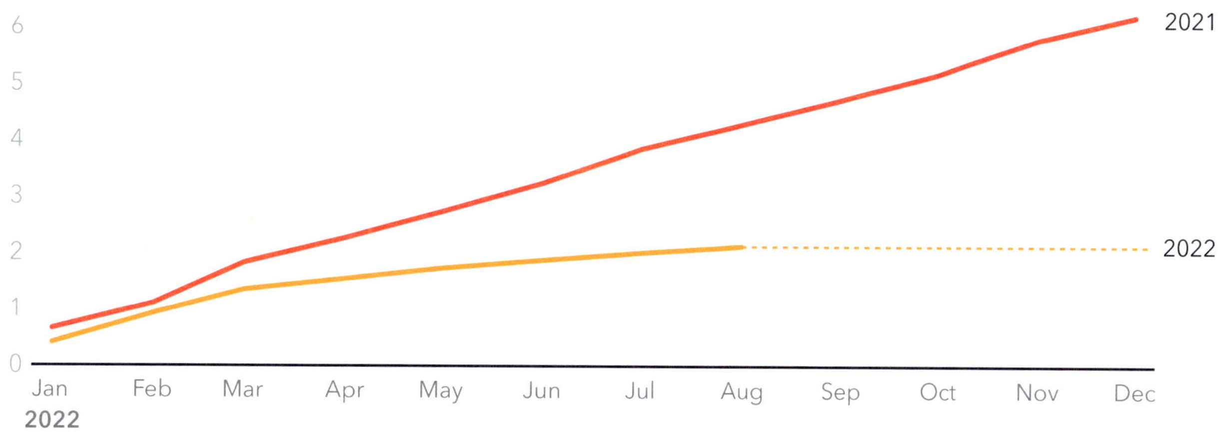

Note: January through August. Data from 53 reporting importers
Source: COMTRADE

Russia and Belarus are attempting to create alternative pathways to export fertilizers. In mid-February, Belarus President Aleksandr Lukashenko said that Russia would build new port facilities for shipping potash from Belarus. In September, Belarus announced that Russia will allow potash exports through the port of Murmansk.

Impacts on importers

How have sanctions affected importing countries? Russia and Belarus jointly account for about 41% of global potash exports, and their disruption has pushed countries to seek alternative suppliers. Take Brazil, the world's second-largest potash importer after the United States. In the first nine months of 2021, about 20% of Brazil's potash imports came from Belarus (Figure 4). For the same period in 2022, that figure fell by 46%, accounting for less than 10% of Brazil's total potash imports. Much of the decline was offset by imports from Canada (up 37% over last year's January–September levels).

Fizre 4: **Brazil potash imports, 2022 vs. 2021, metric tons**

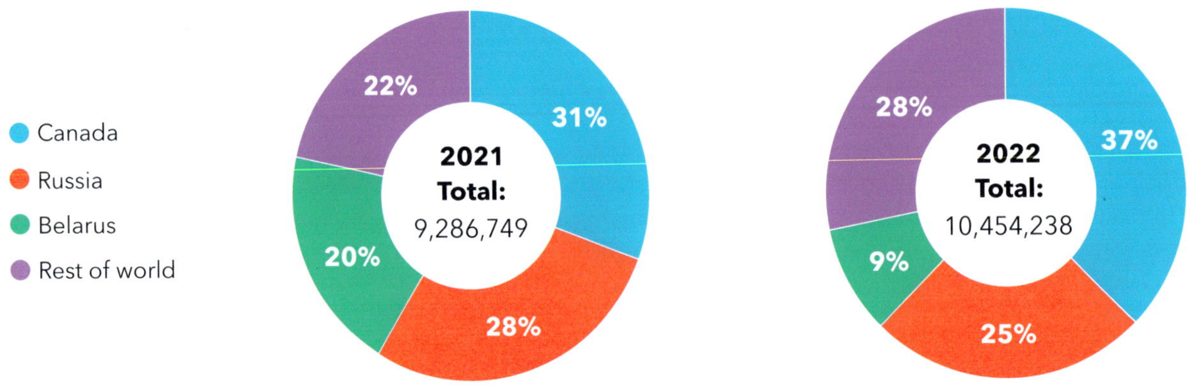

January through September

Source: COMTRADE

Figure 5: **Change in Morocco's anhydrous ammonia imports, 2022 vs. 2021 (Jan.-Aug.), metric tons**

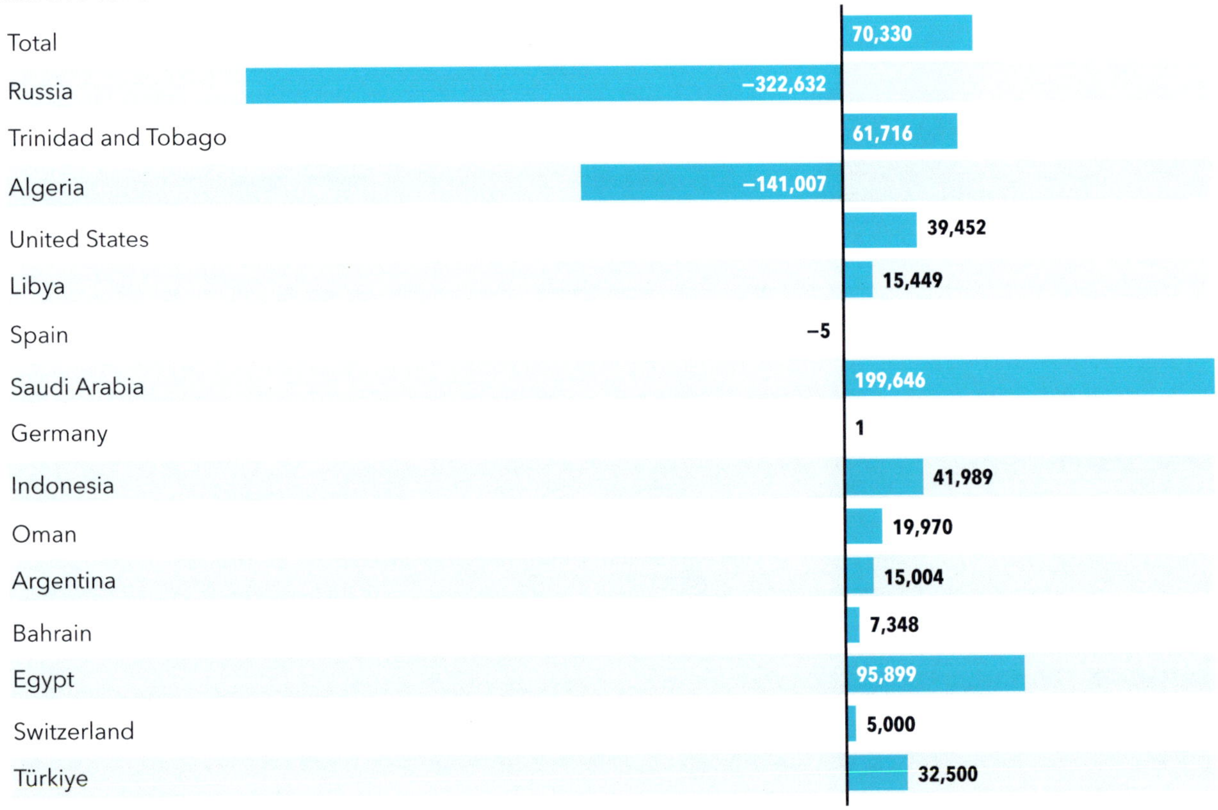

Source: COMTRADE

Russia accounts for about 15% of global nitrogenous fertilizer exports, including products such as ammonia and urea. Importantly, some of these products are part of a global value chain of fertilizer production – thus disruptions in specific inputs could disrupt the production of compound products in other countries. Morocco is the second-largest exporter of phosphate products in the world, in particular diammonium phosphate (DAP) and monoammonium phosphate (MAP), as well as the fourth-largest importer of anhydrous ammonia, traditionally importing from countries such as Russia and Algeria. The sharp reduction in anhydrous ammonia imports from Russia (and Algeria) pushed Morocco to seek a number of alternative suppliers in the first eight months of 2022 (Figure 5), especially sourcing products from Egypt and Saudi Arabia.

Conclusions

Our analysis suggests that carve-outs in trade sanctions appear to have helped sustain exports of some agricultural products and fertilizers from Russia and Belarus, but exports of potash from Belarus and anhydrous ammonia from Russia, which have traditionally utilized EU and Ukrainian port facilities, have fallen significantly. Some affected importing countries have managed to find alternative sources.

What are the prospects for boosting these exports? These issues are central to ongoing negotiations. Russia has told a top United Nations representative that the extension of the Black Sea Grain Initiative depended on the West easing remaining restrictions on Russia's agricultural and fertilizer exports, including re-opening the anhydrous pipeline to Odesa. On November 3, UN Secretary General António Guterres echoed the call for removing obstacles to Russian food and fertilizer exports. The Black Sea grain deal expires November 19, so negotiators may soon produce answers on what happens next.

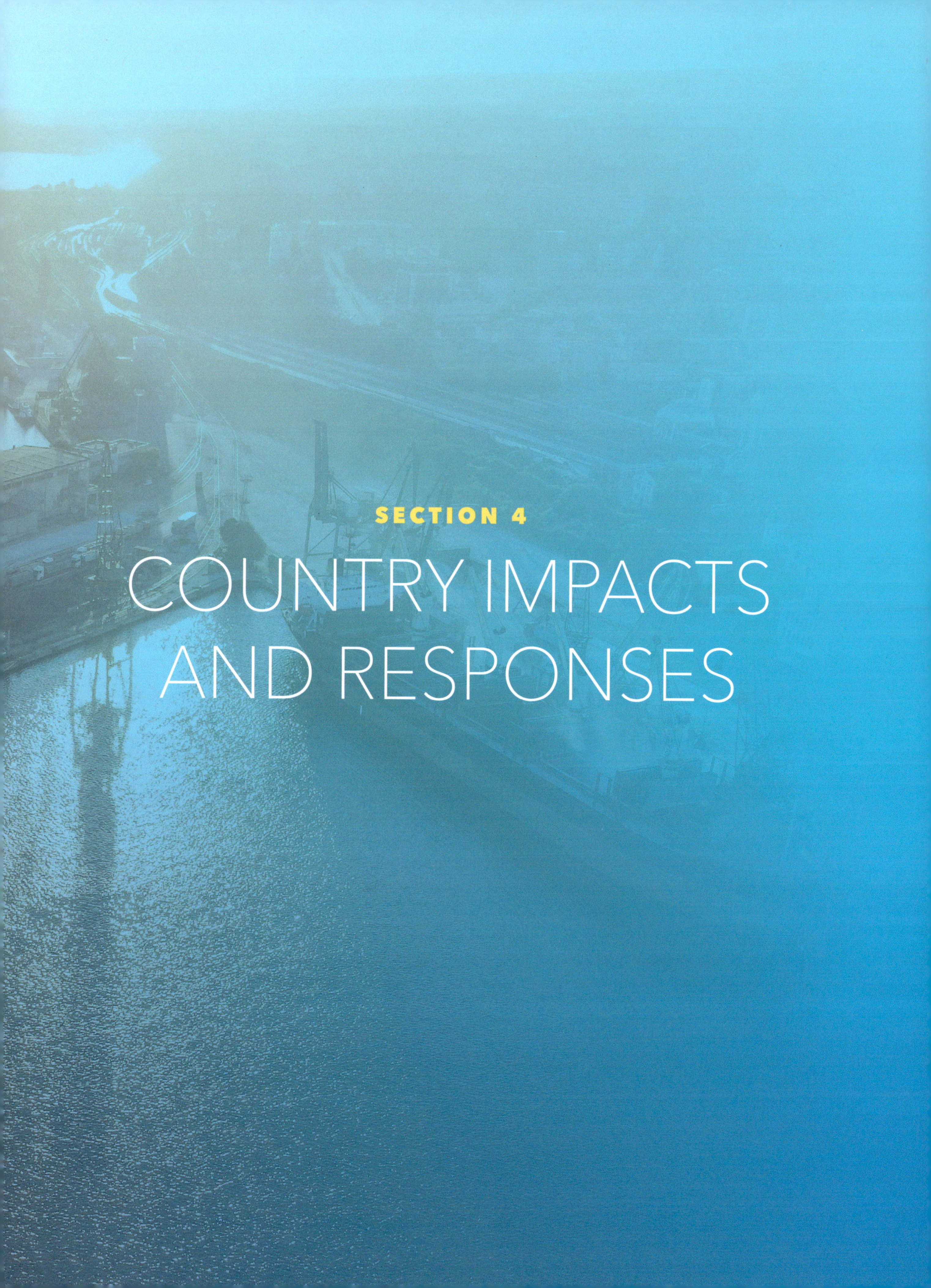

SECTION 4

COUNTRY IMPACTS AND RESPONSES

23 Regional war, global consequences: Mounting damages to Ukraine's agriculture and growing challenges for global food security

Pavlo Martyshev, Oleg Nivievskyi, and Mariia Bogonos

Originally published March 27, 2023.

Russia's all-out war on Ukraine has inflicted devastating impacts that continue to mount more than a year after the invasion. As of September 2022, even before Russia's winter bombing campaign, the total damage to Ukraine's infrastructure was an estimated US$127 billion, equal to 64% of the country's 2021 GDP. More than 14 million Ukrainians have left their homes, including more than 8 million refugees.

These impacts have global economic effects as well – particularly on agricultural markets and food security. Damage to Ukrainian agriculture and production losses also continue to mount. This has created severe economic uncertainty, driving many Ukrainian farmers to the brink of bankruptcy and substantially depressing agricultural output – contributing to high prices and price volatility around the world. In this post, we outline the war's impacts on Ukrainian production and exports of key crops and their continuing global reverberations.

Despite high world prices, farmers in Ukraine suffer huge losses

Over the two decades leading up to the war, Ukraine emerged as an important global supplier of grains and vegetable oil. Grain exports more than quintupled, and the country's share of globally traded grain rose from about 5% to 12–14%. The lion's share of agricultural exports (about 93%) went to destinations in the Middle East and North Africa region and to Europe by sea (via the ports of Kherson, Skadovsk, Berdyansk, Mariupol, Odesa, Bilgorod-Dnistrovskyi, Mykolaiv, Chornomorsk, Olviia, and Pivdennyi).

In the first days of the Russian invasion, Ukraine's Black Sea ports were either occupied or blocked by the Russian naval fleet, contributing to spiking international prices through mid-2022. Large amounts of exportable grains and oilseeds ended up stuck in Ukraine's ports and in inland elevators. Agricultural traders and producers sought alternative routes, including overland by trucks and rail across Ukraine's western borders and through Danube River ports.

These alternative export routes, however, are more expensive and pose various logistical problems, making it difficult to dramatically expand such shipments. For rail, the major bottleneck has been limited loading capacity at railway stations. For trucks, one problem is backups at border crossings. Lines at the border sometimes stretch to more than 50 km and trucks often wait a week or more before crossing, thus driving up export costs.

During the summer of 2022, Ukraine was able to increase its agricultural export shipments by rail only modestly, to about 1 million metric tons per month (Figure 1). Exports by truck have also been limited to about 600,000 tons per month. Thus the total shipment capacity of these alternative routes fell substantially short of demand

and could not bring exports even close to the level of pre-war monthly shipments. This in turn led to oversupplies exhausting existing domestic storage capacities as the 2022 harvest approached.

Due to these problems, export costs surged from the pre-war $30-$40 per ton to $150-$200 per ton, and thus severely depressed domestic grain prices (Figure 2). This predicament led to a sharp decline in revenue of Ukrainian grains and oilseeds farmers – a loss of about $18.5 billion for 2021 and 2022 crops – and drove many to the edge of bankruptcy (Figure 2). The total agricultural losses associated with the war (i.e., forgone revenues and increased production costs) have amounted to $34.25 billion, or close to 75% of Ukraine's agricultural output the previous agricultural year.

Figure 1: Export of grains and oilseeds from Ukraine since Russia's invasion of Ukraine in February 2022, by mode of transportation, million metric tons

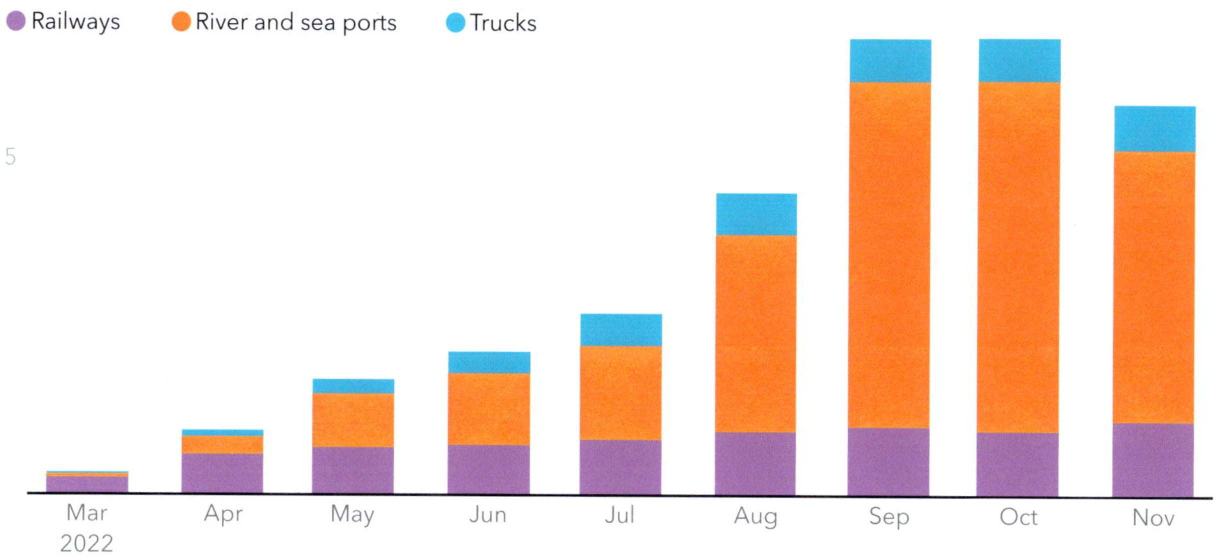

Source: Ministry of Agrarian Policy and Food of Ukraine

So far, limited impact of the Black Sea Grain Initiative

Some relief came with the Black Sea Grain Initiative between Ukraine and Russia, brokered by the United Nations and Türkiye in August 2022, which established a so-called grain corridor from three Black Sea deep water ports (Odesa, Chornomorsk, and Pivdennyi). This resulted in a significant increase in Ukraine's agricultural exports. However, it had only a marginal effect on domestic prices – and farmers' incomes.

When the grain deal took effect, the 2022 harvest had already started. Thus, even as monthly grain exports returned to pre-war levels (Figure 3), supply pressures on the domestic market continued, export costs remained almost at high pre-deal levels, and domestic prices stayed low, without any noticeable movement toward closing the gap with world market prices (Figure 2).

Moreover, the volume of the shipments via the corridor leveled off in October and decreased in November (Figure 3), normally Ukraine's high export season. Contributing to this trend was Russia's repeated questioning of the grain deal, including President Vladimir Putin's accusations, without evidence, of possible arms being masked as grain shipments – frictions that undermined the security of the grain corridor. The dispute flared further when Russia briefly suspended its participation in the grain deal, though it quickly reversed course and the agreement was extended for another 120 days. (On March 18, the agreement was extended, but for how long is disputed. Ukraine contends that the agreement is for an additional 120 days while Russia says that it only approved a 60-day extension).

Figure 2: **Grain prices in Ukraine and EU before and after the Russian invasion**

A. Wheat prices, US dollars per metric ton

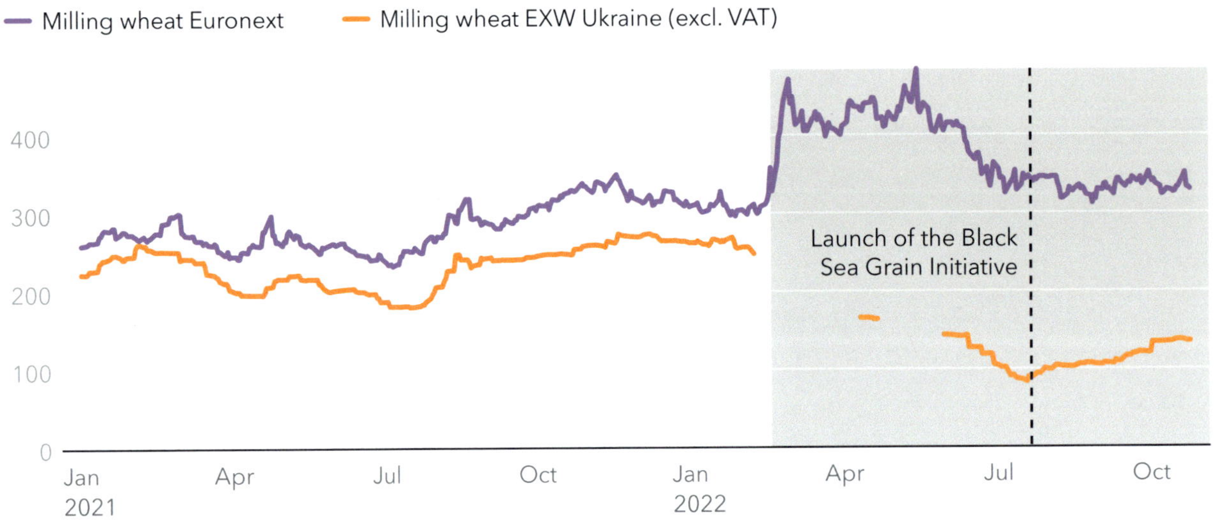

B. Corn prices, US dollars per metric ton

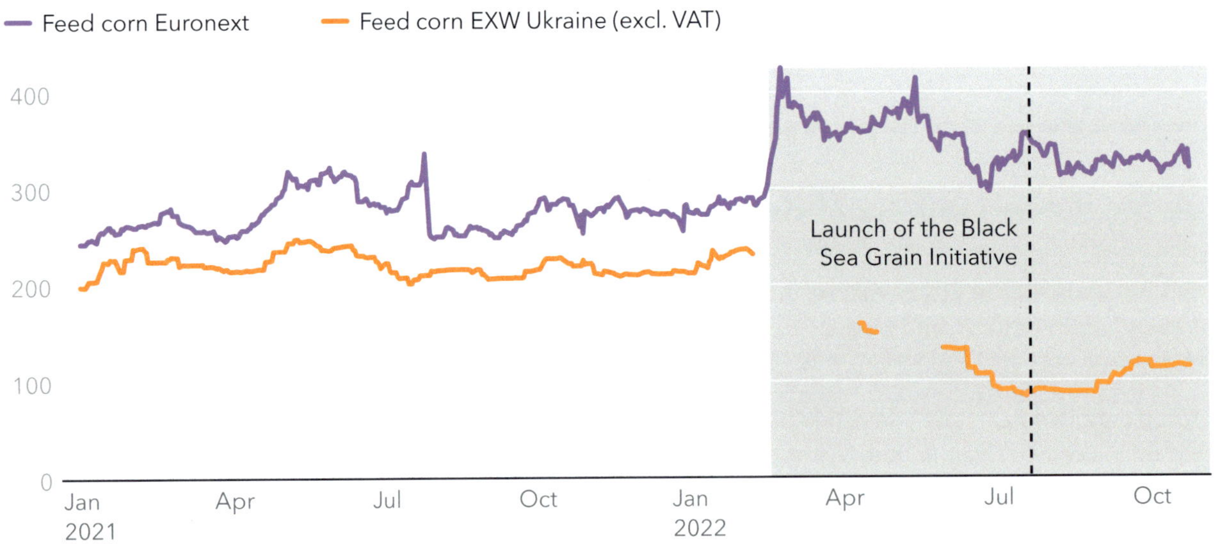

Source: Authors calculations based on data from UkrAgroConsult

Surprises in the structure of exports

Despite these hardships for Ukraine's producers, corn remains the dominant export crop due to large stocks and proximity to main export destinations (primarily the European Union). The share of wheat among overall exports has been increasing gradually since the start of new season in July 2022 (Figure 3). Meanwhile, exports of sunflower seeds have been relatively high (not typical for Ukraine), thanks to high export prices. Usually, more than 95% of the seeds are crushed and exported as oil and meal, but selling to local crushers was less profitable than exporting seeds, even with high transportation costs and rail congestion along the western border.

Figure 3: Export of grains and oilseeds from Ukraine, million metric tons

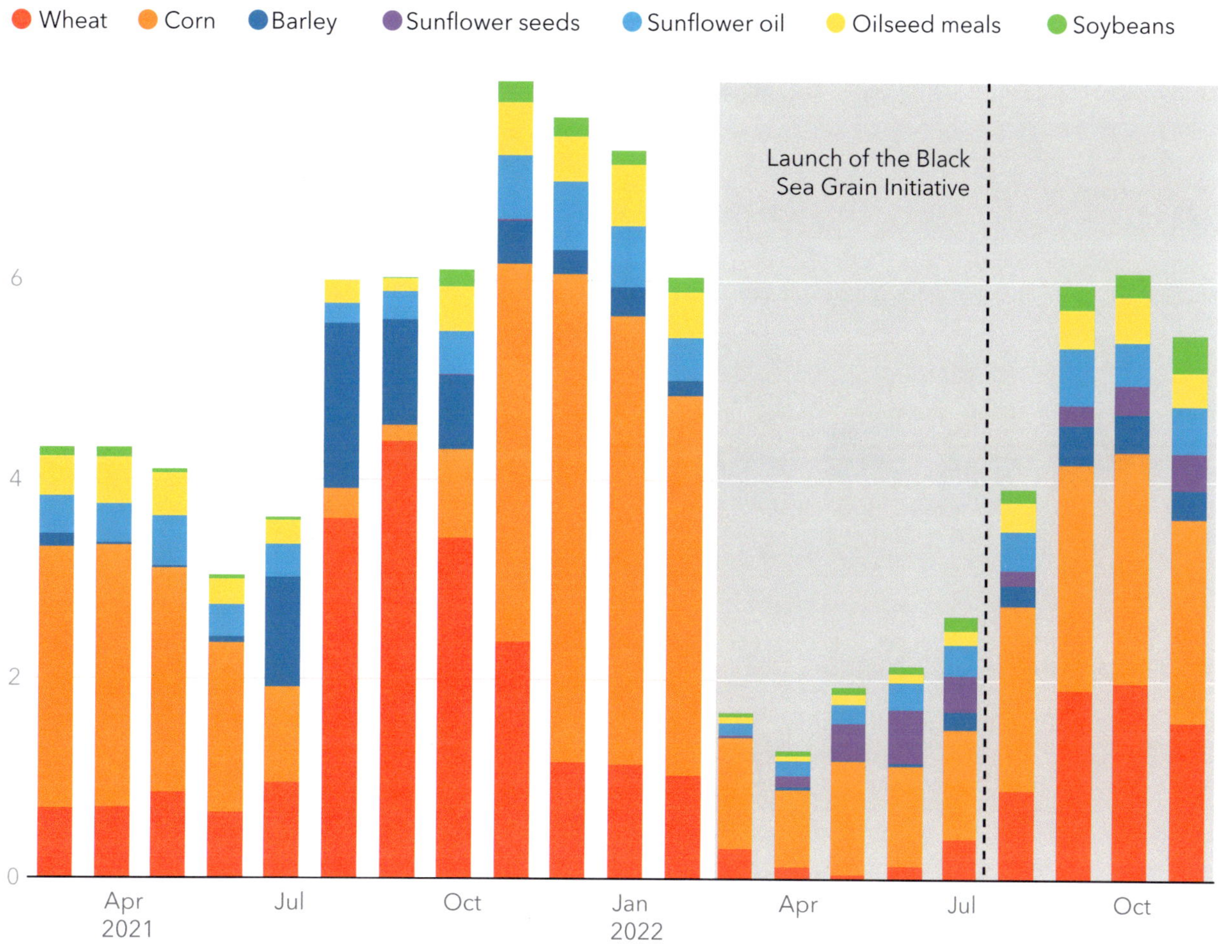

Source: State Statistics Service of Ukraine

Future perspectives and continued threats to global food security

As Figure 4 shows, the U.S. Department of Agriculture expects Ukraine to produce about 75 million tons of wheat, barley, corn, sunflower seeds, rapeseeds, and soya in the 2022/23 season – 31% below the previous year's level. The Ukrainian Grain Association (UGA) and the Ukraine Ministry of Agrarian Policy and Food are even less optimistic, forecasting exports of only 63 million tons. Moreover, due to the rainy fall and huge liquidity problems, the 2022 harvest has been delayed. As of December 1, Ukrainian farmers had harvested only slightly more than 50% of 2022 grains and planted only about 40% of the prior year's winter crop areas, where typically most grain would be harvested and winter crops sown by that time.

Unfortunately, these figures suggest that even the low projections for the current and next season may be optimistic. This is bad news for a tenuous global food security situation. Given current global supply and demand conditions, it is likely that global markets will remain tight and 2022/23 global stocks will drop. Meanwhile, the capacity to replace the expected missing exports from Ukraine in the world are limited, at least in the short run. In other words, it is not an exaggeration to say that every ton of grain from Ukraine will count!

Figure 4: Production of grains and oilseeds in Ukraine, million metric tons

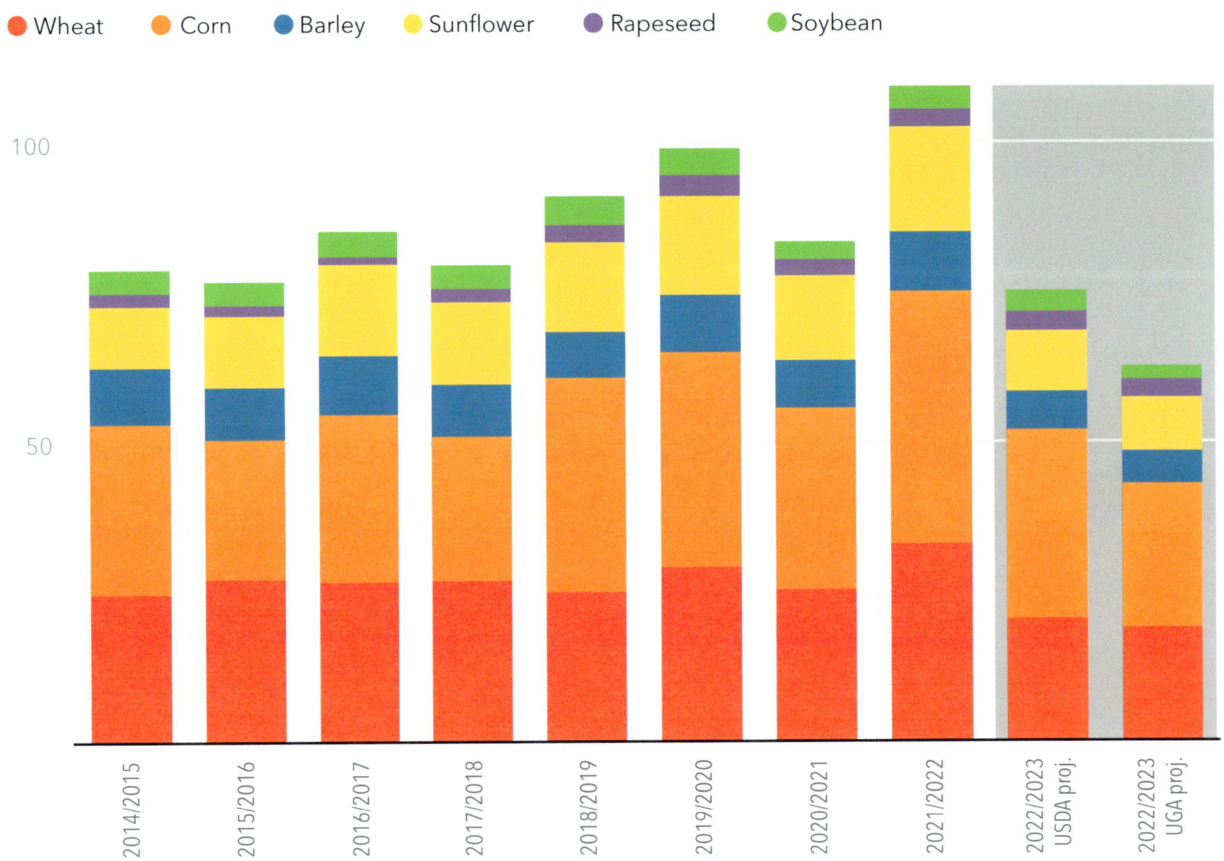

Note: The gray area indicates the projected volume of grains and oilseeds crop in the 2022/23 season.
Source: USDA/FAS PSD database; Ukraine Grain Association

24 The Russia-Ukraine crisis poses a serious food security threat for Egypt

Kibrom Abay, Lina Abdelfattah, Clemens Breisinger, Joseph Glauber, and David Laborde

Originally published March 14, 2022.

Russia's invasion of Ukraine has imperiled global food security – creating suffering within Ukraine and displacing millions while disrupting agricultural production and trade from one of the world's major exporting regions. The latter threatens to drive rising food prices still higher and create scarcity, especially for regions most dependent on exports from Russia and Ukraine – particularly the Middle East and North Africa.

In this post, we consider the impacts of the war on the wheat market, focusing on Egypt. Wheat is a key food item for this country, representing between 35% and 39% of caloric intake per person in the past few years. Wheat imports usually account for about 62% of total wheat use in the country. We conclude by listing a number of key policy actions aimed at diversifying imports in the short term and helping Egypt's agrifood system transformation become fairer and more resilient. The latter is an absolute necessity in the context of looming threats from climate change and water scarcity.

Egypt's vulnerability to Black Sea import disruptions

Egypt is the world's largest importer of wheat. It imports a total of 12-13 million metric tons annually. With a population of 105 million growing at a rate of 1.9% a year, Egypt has become increasingly dependent on imports to meet food needs. Imports of cereal crops have been steadily increasing over the past three decades at a rate higher than that of domestic production (Figure 1).

Figure 1: Egypt's wheat imports, by exporter, million metric tons

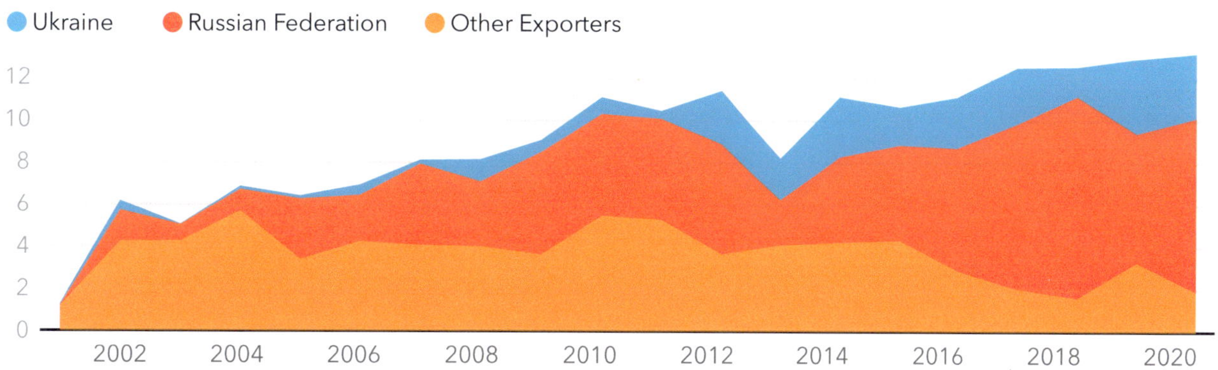

Source: COMTRADE

Egypt's wheat market and trade regime is largely controlled by government agencies. The General Authority for Supply Commodities (GASC), operating under the Ministry of Supply and Internal Trade (MoSIT), handles the purchase of local and imported wheat. International wheat tenders are released every few weeks and the GASC usually handles about half of the total wheat imported, while private trading companies handle the other half.

Despite the government's efforts following the global food crisis in 2007/08 to diversify sources of cereal imports, the vast majority of the increase in cereal imports came from Russia and Ukraine (Figure 1).

The first quarter is the time of year when Egyptian wheat imports are usually at their peak, before the domestic harvest begins in April (Figure 2). Black Sea exporters are also particularly active this time of year, making the current crisis especially disruptive for Egypt.

The GASC is already feeling the impact of the war, which has led to recent cancelation of tenders due to lack of offers, in particular from Ukraine and Russia. Still, there is no fear of shortage in the coming weeks. In early February, Egyptian MoSit Minister Aly Moselhy said that the country held sufficient inventories in its strategic reserve to cover five months of consumption, but the outlook beyond is less clear. With the abrupt closure of Ukraine's ports and current maritime trade in the Black Sea, Egypt will have to find new suppliers if Ukraine is unable to export wheat this year and if sanctions against Russia impede food trade indirectly.

Such opportunities are, unfortunately for Egypt, limited. Currently, wheat producers in South America – Argentina in particular – have larger than usual surpluses from the last harvest available to export. Overall, however, it will be difficult to expand the global wheat supply in the short run. About 95% of the wheat produced in the European Union and about 85% of that in the United States is planted in the fall, leaving those regions little room for expanding production in the near term.

In addition, wheat competes with crops such as maize, soybeans, rapeseed, and cotton, all of which are also seeing record high prices. In combination with record-high fertilizer prices (also exacerbated by the Russia-Ukraine conflict), farmers in some regions such as the United States may favor less fertilizer-intensive crops such as soybeans.

About 20% of world wheat exports come from the Southern Hemisphere (primarily Argentina and Australia), which typically ship in late fall and early winter. In addition, Canada and Kazakhstan are large producers that harvest in the fall. Over the coming year and beyond, their exports may be able to make up much of the deficit created by the loss from Ukrainian production – but at a higher cost due to longer shipping routes and increased transportation costs triggered by higher oil prices.

Figure 2: Egypt's annual wheat imports by quarter and origin, 2017-2020

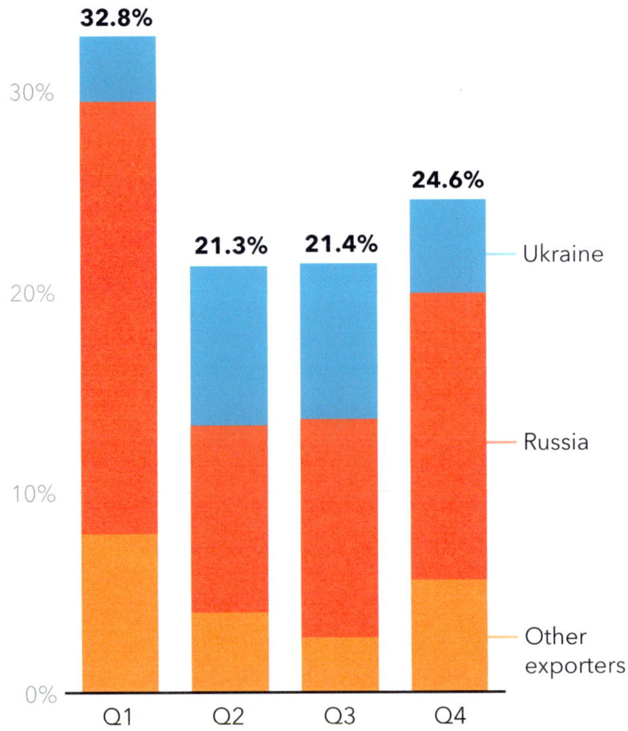

Source: COMTRADE

Rising prices: From global market to Egyptian households

Rising global wheat prices hit a 10-year high at $523 per ton on March 7, 2022 (Figure 3). This is a serious problem for the Egyptian government's budget and a potential threat to consumer purchasing power. Global wheat markets have been tight since mid-2021, predating the war; then the invasion triggered a surge in wheat futures contracts exceeding 50%.

Figure 3: Global wheat prices, nearby future contract, US dollars per metric ton

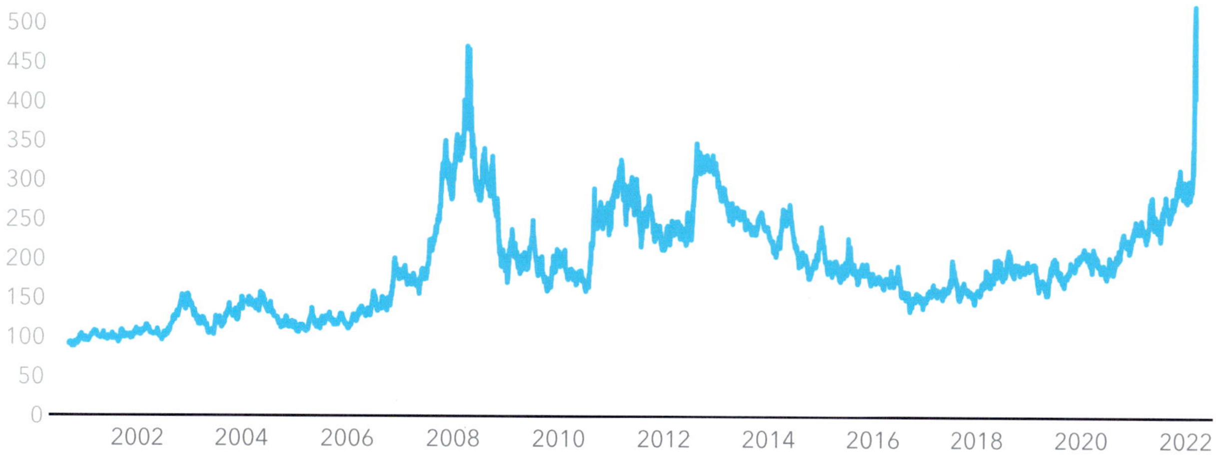

Source: CBOT

Some countries have already imposed export restrictions in response to rising prices, as many did during the food price spikes in 2007/08 and 2010/11 (apart from Russia's export tax regime put in place in 2021). Moldova, Serbia, and Hungary have imposed export bans on some grains, and Indonesia is tightening controls over shipments. These trends, coupled with disruptions in Russia's and Ukraine's exports, will likely put further upward pressures on prices going forward. Even under the most optimistic assumptions, global wheat prices will remain high throughout 2022 and the trend is likely to persist through 2023, given limits on expanding production.

The Egyptian government has been spending about US$3 billion annually for wheat imports. The recent price increase could nearly double that to $5.7 billion. This, in turn, threatens Egypt's baladi bread subsidy program, which provides millions of beneficiaries with 150 loaves of this subsidized bread per month, with about 90% of the production cost borne by the government at an annual cost of $3.24 billion. The program requires about 9 million tons of wheat annually – about half of the total wheat consumption in Egypt and three-quarters of Egypt's wheat imports.

Even just before the outbreak of the Russia-Ukraine war, prices of commodities in Egypt were increasing, with overall annual inflation and food inflation reaching 7.3% and 8.4%, respectively, in January. The war has started adding further pressure to food prices in Egypt, and consumers are feeling these impacts, with annual inflation jumping to a 31-month high in February (8.8%), primarily driven by a surge in food prices (17.6%). While consumers are expected to bear the brunt of the surge in food prices, farmers in Egypt may see some benefits from higher prices because the MoSIT usually sets the price it pays farmers for domestic wheat at a level comparable to prevailing international market prices.

Policy options for a wheat-secure future

In the short term, Egypt needs to diversify its food import sources. The GASC is actively exploring this option, while also increasing planned procurement from domestic sources by 38% over last year's figure. Overall domestic production for the marketing year (MY) 2021/22 is projected to be 9 million metric tons, only 1.12% higher than the previous year's 8.9 million tons. In addition, the government has decided to ban exports of staple foods, including wheat, for three months to limit pressure on existing reserves. These measures could alleviate some pressures on the Egyptian economy, but they will likely affect Egypt's longer-term trade relations, and even shift the burden of the crisis to neighboring countries with limited capacity to manage the crisis. For example, export restrictions that limit Egyptian food exports to Yemen could lead to drastic humanitarian consequences in this country.

In the long term, Egypt needs to explore options for reducing the gap between domestic supply and demand. Here are some of its options.

Boosting domestic wheat production will be challenging, as Egyptian farmers are already achieving high yields, relying on high input and water use. While there are some opportunities to expand arable land, modernize farming systems, and improve water management practices, the country's principal focus should be to adapt the farming system to address imminent water shortages and climate change threats and increase resilience, rather than unsustainably expanding production.

Reducing the high consumption and waste of bread has significant potential. Egyptians on average consume about 145 kg of wheat per capita annually – double the global average. Reductions in wheat consumption and food waste can serve the dual purpose of improving Egypt's self-sufficiency while also addressing malnutrition by shifting consumption from wheat to a more diverse set of food groups. Egypt has high rates of overweight and obesity, which are linked to food subsidies and associated consumption of energy-dense foods.

Improving the efficiency and targeting of Egypt's Tamween food subsidy program, which provides beneficiaries with ration cards for various foods, also holds promise. The program absorbs a large share of imported wheat and vegetable oils. Reforming it could reduce inefficiencies in the wheat sector and the cost of running the program. Improving the targeting of Tamween could also reduce government spending, given that nonpoor households receive about two-thirds of the total value of food subsidies, while three-quarters of those households benefit in some form from the program. IFPRI's work shows that improving the targeting of Tamween can enhance the welfare of poor households and cost savings from these reforms can be channeled to other targeted food security and nutrition interventions.

In conclusion, the Russia-Ukraine war poses a big challenge to global food security and particularly difficult obstacles for Egypt. The country implemented significant economic reforms in 2016, which together with a massive stimulus package helped it to manage the COVID-19 pandemic. Now, just as it embarks on post-COVID recovery and the second phase of the economic reform program, another crisis has struck that seriously threatens food security. The short-term and long-term impacts will of course depend on how the war unfolds and affects exports from Russia and Ukraine over the coming months and years. Impacts on Egypt will also depend on other countries' responses to global price hikes and cereal shortages. Egypt can mitigate some of these impacts with short-term actions as outlined above, but major global shocks like the Russia-Ukraine war are also reminders of the need for longer-term reforms and solutions.

The IFPRI-Egypt program website has published this blog post and its figures translated into Arabic on their website. Please click here for the Arabic version.

25 Food price shocks and diets among poor households in Egypt

Kibrom Abay, Naureen Karachiwalla, Sikandra Kurdi, and Yousra Salama

Originally published December 29, 2022.

Because of its dependence on food imports, Egypt is particularly vulnerable to the high world food prices and trade shocks triggered by Russia's February 2022 invasion of Ukraine. Imports account for more than 40% of Egypt's calorie consumption, and the country is the world's largest importer of wheat, with about 85% coming from Russia and Ukraine before the war began. Since then, supply disruptions, high global market prices, and other factors have led to sharp rises in domestic food inflation, which reached 31% in November. This in turn is a major driver of rising overall annual inflation, which soared from 6% to 19% between January and November 2022 – the highest rate in five years.

In this post, based on a recent survey, we assess how Egypt's poor households are responding to these economic stresses, and how the Egyptian government's social protection efforts have affected consumption since the conflict began. In general, the survey indicates that many poor households cut back on consumption of some unsubsidized nutritious foods while consumption of subsidized foods remained unaffected, suggesting the national food subsidy program plays an important role.

Annual inflation rates for some food groups are even higher than the 31% overall figure (Figure 1). Most notably, the inflation rate for bread and cereals approximately quadrupled during the conflict.

Figure 1: Annual food inflation rates before and after the Russia-Ukraine crisis

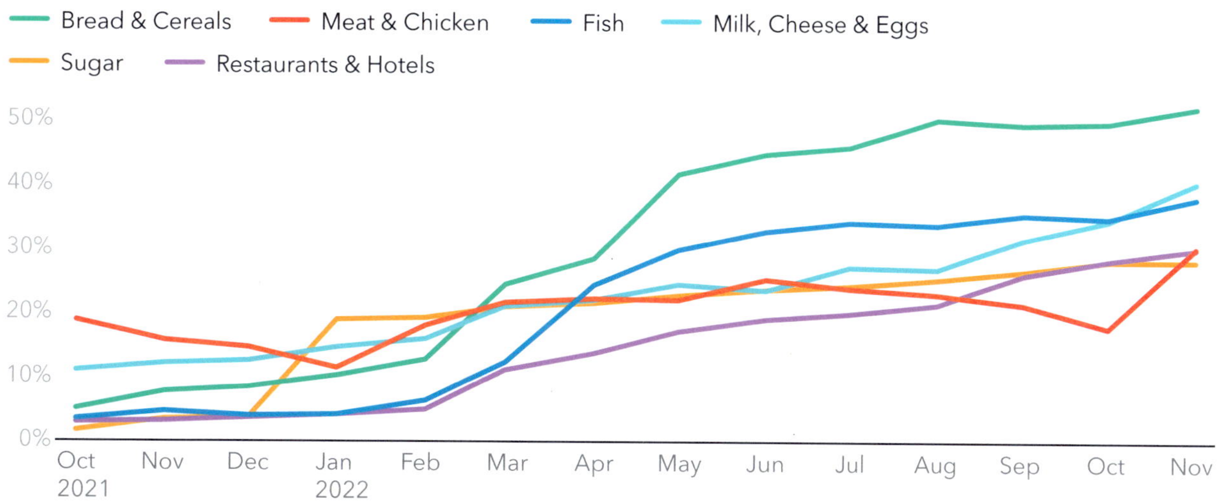

Note: For each month listed, the comparison is to that same month in the previous year
Source: IFPRI 2022 Phone survey data

Poor households are the most vulnerable to food price inflation, mainly because they allocate a larger share of their income to food consumption. However, strong social protection infrastructure may support vulnerable households during food price shocks. Egypt makes significant investments in social protection and food subsidy programs. For example, the national food subsidy program, Tamween, reaches about 73% of households (MSIT 2019). Throughout the conflict, Egypt's government has continued to subsidize key food items, keeping their prices unchanged. Consequently, we can expect households to respond differentially to subsidized and unsubsidized foods.

IFPRI surveyed more than 6,000 poor and near-poor households from across Egypt by phone in October and November 2022. Households were asked to report how, since March 2022, they perceived that their consumption habits changed, reasons for changes in consumption, perceived changes in prices, and coping strategies that they employed to respond to the price shocks.

Figure 2 displays the changes in consumption for foods for which the largest proportion of households reported either staying the same or decreasing. Overall, across 21 food items, households reported a decrease in consumption of unsubsidized foods and either a slight increase or the same amount of consumption of subsidized foods. Many respondents reported reducing consumption of unsubsidized food items. For example, 85% of households reported reducing meat consumption and 75% reported decreasing consumption of chicken and eggs, which makes intuitive sense given large price increases for these items. Unfortunately, these are nutritious food groups that provide key sources of protein, suggesting a potential decrease in diet quality. Fish and milk are also important sources of protein and other nutrients whose prices increased substantially.

Figure 2: Reported changes in consumption since March 2022

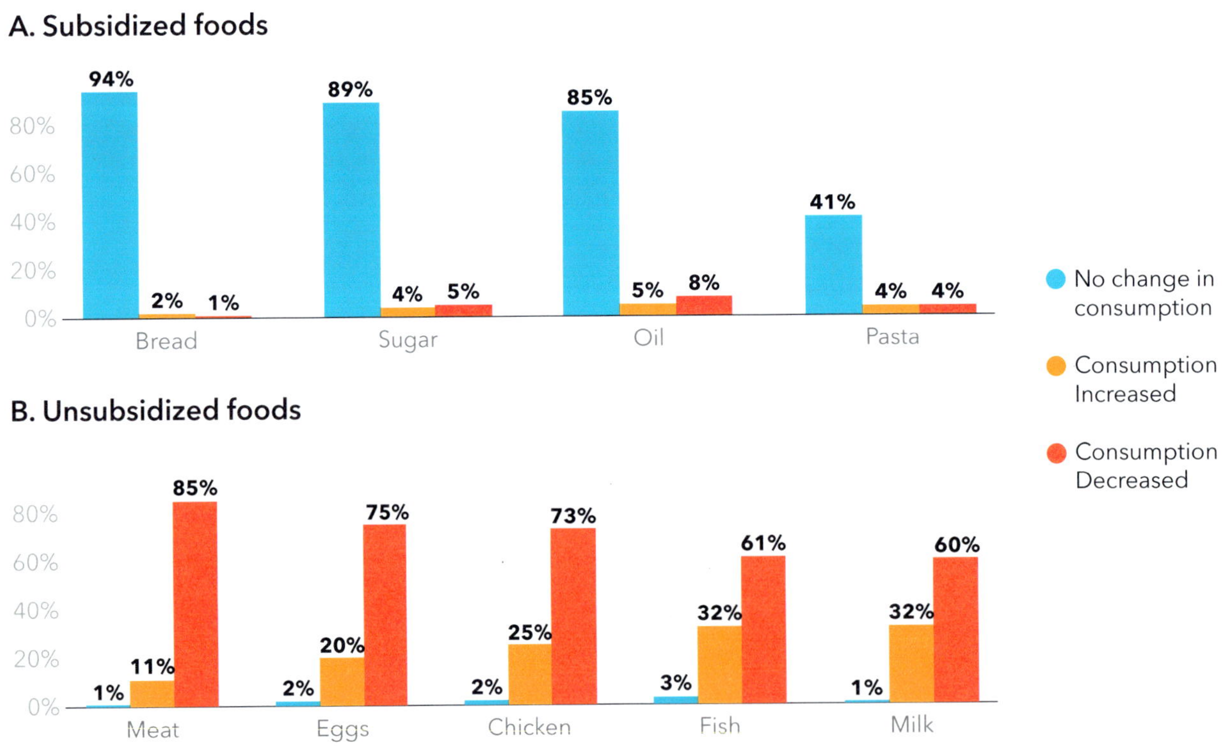

Source: IFPRI 2022 Phone survey data

Conversely, some households reported increasing their consumption of less nutritious food groups to reduce food spending. There were only two foods for which households reported an increase in consumption. Household consumption of potatoes and unsubsidized types of pasta increased among 21% and 14% of households, respectively. These are starchy food items that, while filling, tend to have fewer nutrients. Figure 2 also shows that the reported consumption of the major subsidized foods (bread, sugar, and oil) was unchanged for most households. However, 24% of households reported that their consumption of unsubsidized types of bread decreased.

These patterns suggest a potential switch from unsubsidized to subsidized products. Note, importantly, that these changes could have occurred at any time during the period since March 2022, so the scale of the effects partly reflects this long period of time during which, at any point, households may have reduced their consumption of various foods.

Households were also asked about the reasons for reducing consumption of food items. More than two-thirds reported that price increases were the main cause. More than 70% of households reporting a reduction in chicken and egg consumption cited price increases as a main reason, while for those cutting back on meat consumption, 85% cited price increases (Figure 3). This is consistent with commodity-specific inflation rates shown in Figure 1. However, it is worth noting that these are perceived changes and actual consumption patterns may evolve differently, as suggested by a recent IFPRI study. Recent studies comparing perceived effects of COVID-19 and actual consumption reductions show that perceived changes are larger than actual reductions in consumption. Our further analysis will explore this as well as the role of alternative social protection programs in cushioning the adverse impacts of food inflation.

Figure 3: Reasons for reduction in consumption

Meat

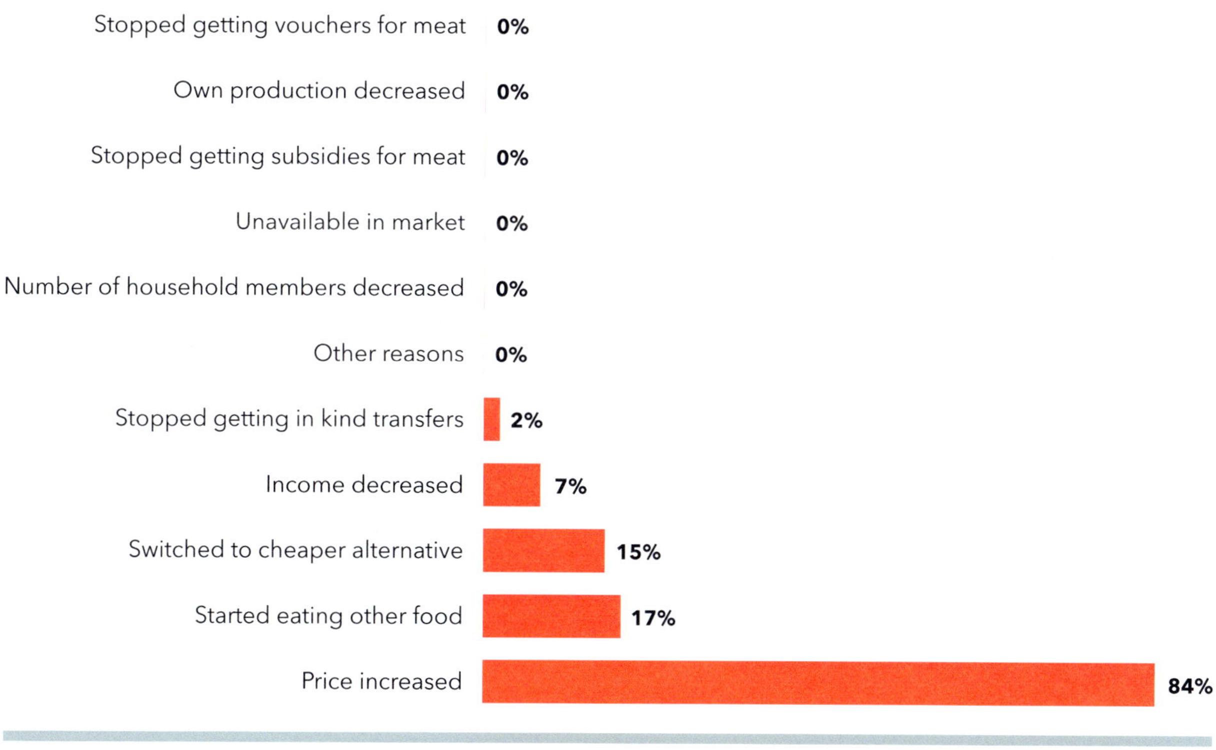

Figure 3: Continued

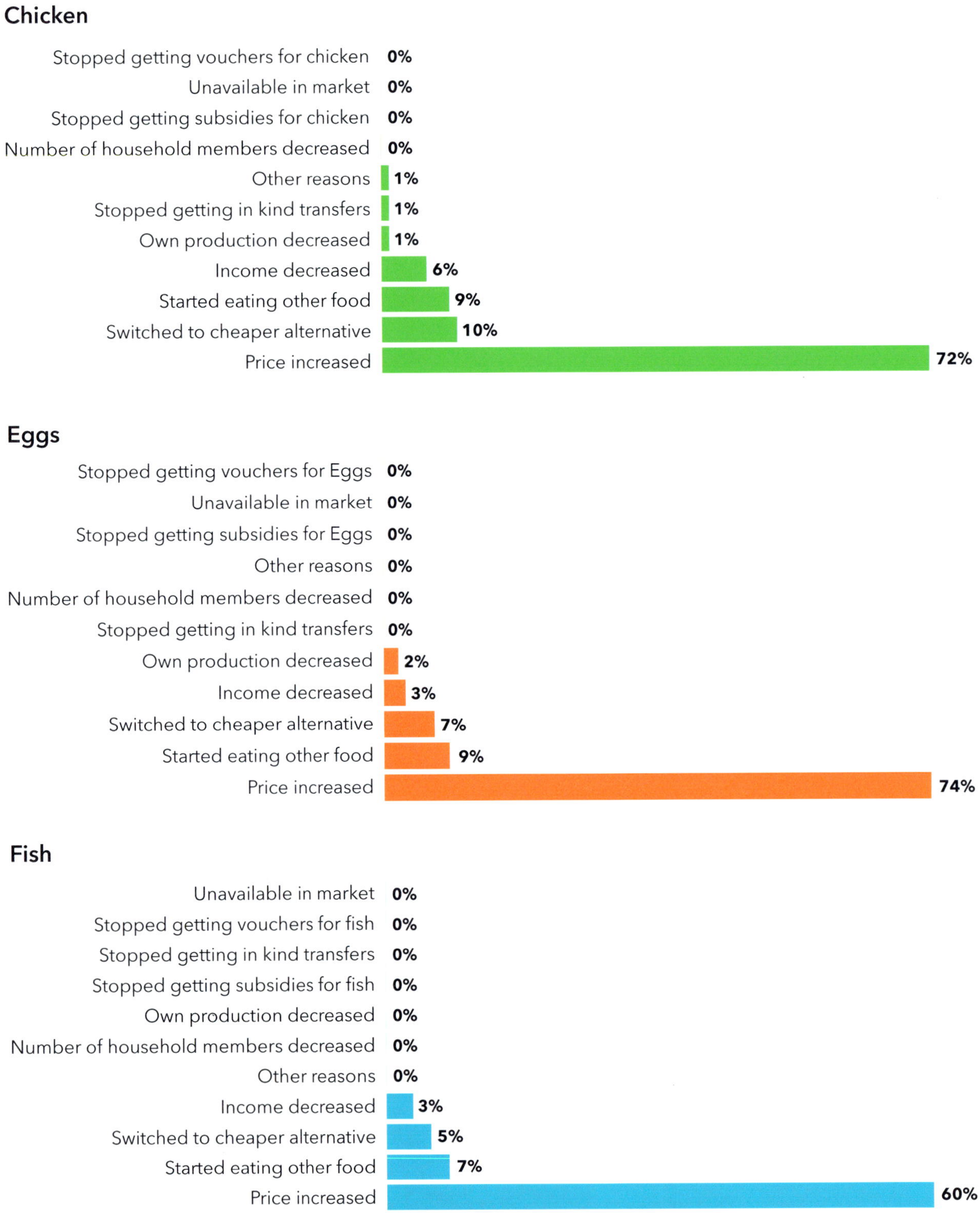

Finally, households were asked to report how they had responded to the food price shocks (Figure 4). Some of these coping strategies are likely to have lasting adverse impacts on livelihoods. For instance, selling productive assets or reducing investments in education and health can adversely affect human capital accumulation and reduce future earnings. The most frequently used coping strategy was stopping the repayment of debts (84%), followed by shifting to lower-quality foods and brands (70%), and reducing food consumption (47%). A substantial share of households also reported reducing expenses on health (43%) and education (25%), which may have important implications for future well-being.

Figure 4: Coping strategies employed since March 2022

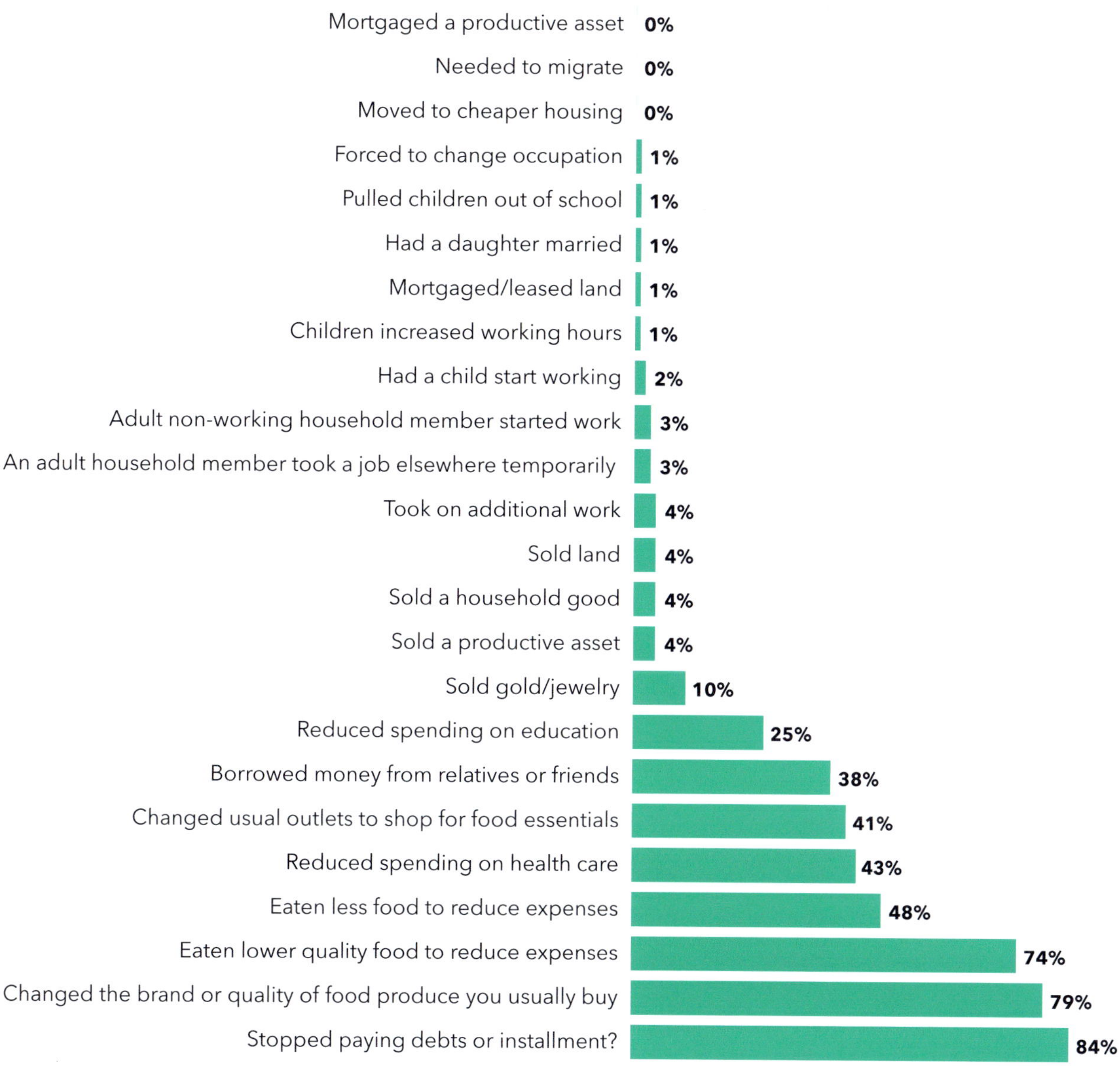

Note: Multiple choices allowed per household

Source: IFPRI 2022 Phone survey data

Overall, despite the significant increase in price of many staple food items, including bread and cereals, these results suggest the government food subsidy program may have at least partly protected households' consumption of these food items. The shift away from high-protein foods and toward less nutritious foods is, while expected, unfortunate and may contribute to the double burden of malnutrition and exacerbate the already high rates of overweight and obesity in Egypt, which are linked to food subsidies and associated consumption of energy-dense foods. Ultimately, poor Egyptian households faced substantial difficulties in 2022 that will set many back from graduating from poverty.

This work is part of the CGIAR Research Initiatives on National Policies and Strategies (NPS). CGIAR launched NPS with national and international partners to build policy coherence, respond to policy demands and crises, and integrate policy tools at national and subnational levels in six countries in Africa, Asia, and Latin America. CGIAR centers participating in NPS are The Alliance of Bioversity International and the International Center for Tropical Agriculture (Alliance Bioversity-CIAT), International Food Policy Research Institute (IFPRI), International Livestock Research Institute (ILRI), International Water Management Institute (IWMI), International Potato Center (CIP), International Institute of Tropical Agriculture (IITA), and WorldFish. We would like to thank all funders who supported this research through their contributions to the CGIAR Trust Fund.

26 One of the world's worst economic collapses, now compounded by the Ukraine crisis: What's next for Lebanon?

Clemens Breisinger, Nadim Khouri, Joseph Glauber, and David Laborde

Originally published May 6, 2022.

High food prices and supply disruptions triggered by the Russia-Ukraine war are hitting Middle Eastern and North African (MENA) countries like Egypt, Sudan, and Yemen hard, partly due their heavy dependence on wheat imports. But in the region, Lebanon – already in the midst of one of the world's worst economic collapses since the 1850s – is uniquely vulnerable to food security impacts from the Russia-Ukraine conflict.

A recent World Bank report calls Lebanon's current crisis "The Great Denial" – referring to an ongoing breakdown of government services, civil society, and the economy.

Over the past decade, the country of 6.8 million has seen one crisis after another. From 2011 onward, the Syrian civil war sent 1.5 million refugees across the border. Starting in 2019, the collapse of the immense Ponzi scheme at the heart of Lebanon's banking sector led to the steady decline of the Lebanese pound.[1] Then came the economic stagnation of the COVID-19 pandemic, and then the devastating double explosion at the port of Beirut on August 4, 2020. In addition to killing more than 200 and injuring 7,000, the explosions destroyed businesses and the port infrastructure, including immense wheat silos. As a result of these multiple crises, 61% of people in Lebanon reported challenges in accessing food and other basic needs at the end of 2021, compared with 41% at the end of 2020, according to the World Food Program (WFP).

Now, the war in Ukraine is almost certainly driving that figure higher. In this post, we look at the continuing toll of Lebanon's economic crisis, how impacts from the Russia-Ukraine conflict are further compounding food insecurity, and some potential policy responses.

Economic crisis

Lebanon's current financial crisis is almost unprecedented in its severity. While the last year has been particularly dreadful in terms of economic stability, the country has not recorded economic growth since 2017. Its GDP in constant US dollars has declined at an alarming rate: −1.7% in 2018, −7.1 % in 2019, and −22% in 2020. At the same time, the large influx of Syrian refugees led to a population increase of 30% in 10 years. Thus GDP per capita has declined dramatically: adjusted for purchasing power parity, it fell from $16,000 in 2017 to $12,100 in 2020, a drop of 25%, with even more pessimistic estimates for 2021 and 2022. In this context, poverty has surged – more than three-quarters of Lebanon's population was estimated to live in poverty at the end of 2021, including one-third below the extreme poverty line.

1 It is still officially pegged to the US dollar at a rate of LL1,500 per USD but now hovers around LL24,000 in the black market. There are various exchange rates formally recognized by the Government (https://www.bdl.gov.lb/news/more/8/564/21).

Figure 1: Applied exchange rate: Lebanese pounds per US dollar

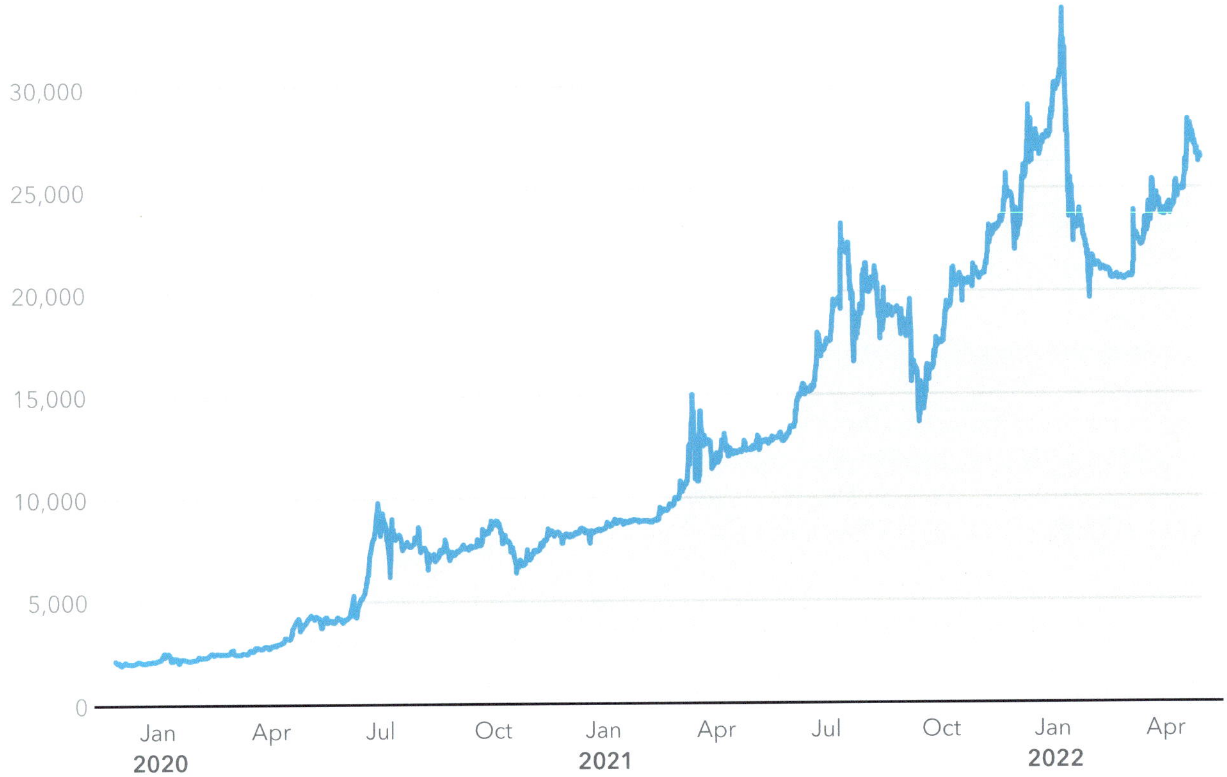

Note: Exchange rate based on market prices. Officially, the Lebanese pound is pegged to the USD at 1 USD = 1,507.5 LBP
Source: Lirarate.org

Meanwhile, government revenue plunged by about 40% in four years (in constant US dollars) – limiting its capacity to respond to the crisis and forcing it to rely on domestic or external debt to act. At the same time, public trust in the country's monetary authority has evaporated; the current governor of the Central Bank of Lebanon is targeted by numerous corruption investigations.

As a result of all these factors, trust in the Lebanese currency has vanished. The effective exchange rate has collapsed by nearly 95% in two years (Figure 1) and domestic inflation has skyrocketed: annual consumer price inflation in February 2022 reached 215%. Food price inflation is nearly double that figure at 396%.

Beyond food products, even day-to-day consumption goods have become luxury items; the cost of a gallon of diesel fuel represents half of the Lebanese monthly minimum wage (675,000 pounds or US$25 at the current exchange rate).

Exchange rate depreciation is a major concern, compounding the rise in global food prices (+34% in 12 months). The collapse of the Lebanese pound means that food imports are actually 180% more expensive in local currency than one year ago.

Figure 2: Daily per capita calorie consumption in Lebanon, by origin and product, average 2018-2020

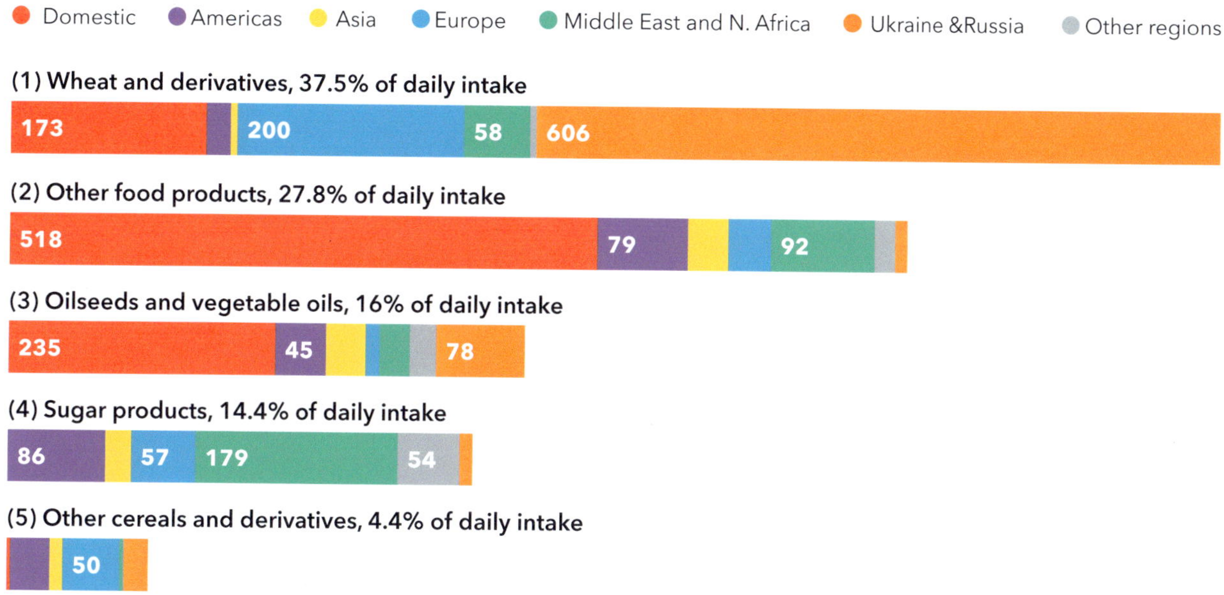

Note: For vegetable oils, the domestic value includes cooking oils processed locally, even if the oilseeds are imported (soybean case)

Source: IFPRI based on FAOSTAT & COMTRADE

Figure 3: Lebanon's wheat imports, million metric tons, by exporter

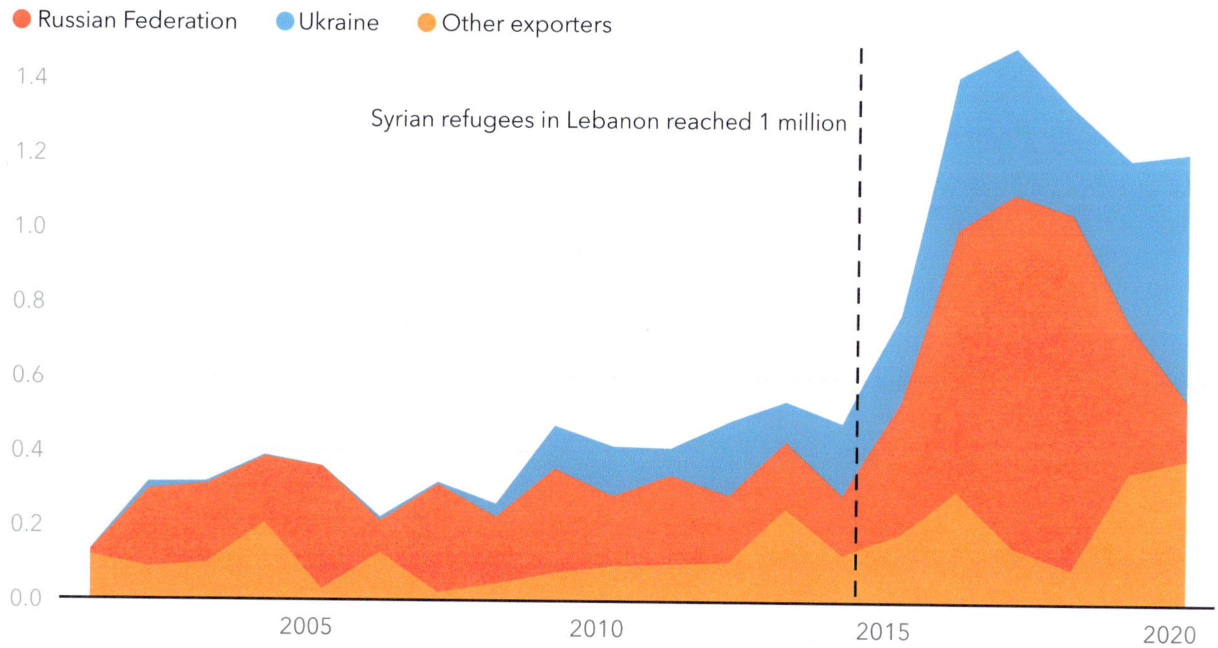

Source: COMTRADE

Why the war in Ukraine matters for Lebanon's food supply

Like other countries in the Eastern Mediterranean, Lebanon depends heavily on food imports from Russia and Ukraine. Over 25% of total calories consumed by the average Lebanese household originate from those two countries, either through wheat-based products or sunflower oil. Wheat accounts for 38% of total calorie consumption and wheat imports over 80% of total wheat consumption. In recent years, between 70% and 90% of imported wheat has originated from Russia and Ukraine (Figure 2).

Wheat imports averaged about 1.2 million metric tons during 2019–2021 – a tenfold increase over 1999–2001 (though the level actually represents a decline since 2017 due to the deteriorating economic situation and declining purchasing power) (Figure 3). Lebanon's wheat storage capacity also significantly declined due to the destruction of silo facilities in the 2020 port explosion. The decision to demolish the damaged silos has just been made and future actions aiming to replace the 100,000 metric ton capacity lost have not yet been initiated. Current Lebanese grain stocks are low, estimated at only 30–40 days.

Lebanon also depends heavily on sunflower oil imports from Russia and Ukraine. Sunflower oil accounts for over two-thirds of vegetable oils consumed by Lebanese households.

Limited options for the Lebanese government

Unlike some other MENA countries such as Egypt, where a government authority (the General Authority for Supply Commodities) procures imports of what are deemed "strategic" commodities such as wheat, until recently Lebanon had a purely commercial practices-based procurement system in place, with millers the key agents and drivers of the supply chain. But in response to skyrocketing inflation, the government began procuring wheat on an intermittent basis to keep bread prices low. This direct involvement has now mushroomed: government outlays represent a reported 90% of the cost of wheat, a total of $390 per ton.

This policy will be difficult to sustain over time. Lebanon was already struggling to buy wheat in March due to lack of available funds, so the government was waiting for the opening of a credit line (for $26 million) by the Central Bank to purchase Indian wheat (50,000 tons, or about one month of consumption). In order to address these foreign exchange shortages, the government recently negotiated a World Bank emergency loan to cover the financing of about six months of wheat imports.

Low wheat prices also encourage smuggling abroad. This means it is crucial to clamp down on contraband wheat and bread, which defeat the purpose of public subsidies and waste already-limited government funds.

Given these problems, Lebanon should consider reverting to a policy of non-intervention in wheat imports as early as possible. As the government prepares to reduce subsidies, it should review and reassess subsidy levels to ensure fairness and reduce waste. In the medium and long terms, the government should be supported in building institutional capacity and a comprehensive social safety net. One of the focuses should be on targeted safety net interventions, including existing programs, in support of equitable food consumption and nutrition security for the poor.

Meanwhile, supplies are tight. To manage the lower availability of wheat and lower prices, the government has tried to deter the use of wheat for so-called luxury goods like cakes and croissants. But during Ramadan, which ended May 1 – with the custom of consuming such sweets for iftar, the evening meal – the measure was highly unpopular.

In addition to these challenges, disruptions in milling operations increased in early April. One factor was that the last shipment received from Ukraine, in December 2021, was damaged by humidity and moisture.

To alleviate these issues, Lebanon is also looking beyond some of its traditional wheat suppliers like India and other sellers such as the United States and Kazakhstan. But this diversification strategy may be hampered by the increasing number of countries enacting export restrictions (for example, Türkiye; follow the evolution of the situation on IFPRI's Food and Fertilizer Export Restrictions Tracker), along with increasing competition from countries like Egypt for wheat imports. A recent World Bank report recommends strengthening the Ministry of Economy and Trade to better monitor international wheat prices and assess offers being made and negotiated for imports into Lebanon.

Lebanon also has some potential to increase its domestic wheat production, particularly for bread. The government should explore supporting such efforts through research and other market-friendly means.

Finally, since the port explosion, "just-in-time" wheat import scheduling has been implemented to ensure that the relatively small-scale storage capacity at mills is not overwhelmed. The discussions surrounding the reconstruction of the silos also offer the country the opportunity to reconsider its storage policy or study the feasibility of assigning a regional vocation to the rebuilt silos, as the entire region could benefit from a coordinated policy on strategic storage of wheat and grains in general. With adequate supplies, mills are also capable of producing beyond national/local needs and could look into the possibility of export to neighboring countries.

Conclusion

The situation in Lebanon shows the many ways macroeconomic conditions are critical in determining how well a country responds to food system shocks. A "perfect storm" of domestic economic and political crises, a large refugee influx, and rising food and energy prices has struck the country. To prevent food insecurity from reaching intolerable levels and triggering significant social unrest, Lebanon does not have the resources to address the crisis systematically, or by itself; it must rely on regional and global support and ad hoc solutions. As part of such immediate support, the International Monetary Fund is currently discussing a macroeconomic stabilization program and the World Bank is discussing a loan to import wheat.

Unfortunately, such ad hoc solutions are not sustainable over the medium or long term. Still, they may be essential to at least make it through some of the immediate deadlines for the country. There are upcoming legislative and presidential elections in succession. Food riots would disturb both. Avoiding this outcome requires a combination of upstream measures such as ensuring wheat supply for the coming few months, and downstream ones such as ensuring equitable distribution of bread (and food in general) through safety net interventions.

These immediate interventions require international solidarity and direct international assistance that should not attempt to interfere with the domestic political process. However, more profound and long-term solutions will need to rely on internal actors and political determination. Many in Lebanon see some hope for more sustainable and effective solutions to the economic crisis in the upcoming opportunity for renewal of the country's leadership. Could a private-public solution to the wheat supply problem, exacerbated by Russia's invasion of Ukraine, show the way to addressing Lebanon's longer-term problems?

27 The Russian invasion of Ukraine threatens to further exacerbate the food insecurity emergency in Yemen

Sikandra Kurdi, Clemens Breisinger, Joseph Glauber, and David Laborde

Originally published March 23, 2022.

The Russia-Ukraine conflict has roiled agricultural markets, particularly the wheat market, which has seen prices rise by 30% since Russia invaded on February 24. This post focuses on the impacts of the crisis on Yemen, whose poverty, civil war, and dependence on wheat imports from Russia and Ukraine make it uniquely vulnerable to the current market and supply disruptions.

Yemen's civil war has ravaged the country since 2015, and lately serious impacts from the COVID-19 pandemic have worsened its already precarious food security situation. The prevalence of undernourishment, as measured by the UN's Food and Agriculture Organization (FAO), exceeds 45%, and more than half of the population relies on some type of in-kind food assistance.

As with Egypt and other countries in the Middle East and North Africa (MENA), a large portion of the Yemeni population's caloric intake consists of wheat products, and the country relies heavily on imports – including wheat and oil – from Russia and Ukraine. The Russia-Ukraine war threatens to exacerbate undernourishment and the reliance on external assistance further by increasing the cost of key food and energy products.

Yemen's vulnerability to higher wheat and energy prices

Named "Arabia Felix" by the ancient Romans for its fertile lands, Yemen was historically self-sufficient in agriculture. But in the 1970s, wheat imports began to rise as the labor force shifted from predominantly agricultural work to migrant work in neighboring Gulf countries during the oil boom, and population growth also accelerated. Other factors contributed to the shift to imports, including a focus on irrigated agriculture reliant on the unsustainable use of pumped ground water, less labor available for maintaining traditional terraces, and increasing commercial cultivation of khat.

The civil war compounded a deteriorating macroeconomic situation: Between 2003–2005 and 2018–2020, total imports rose by over 85% while exports fell by 72% (Figure 1). The total value of imports is now six times that of exports, and food imports alone exceed total exports.

Yemen today relies heavily on remittances and international aid to supplement local income. While crude oil remains the core exported commodity, the country spent US$3.5 billion on average on imported food products and refined oil alone – about 15% of its GDP. This creates a significant vulnerability to rising prices on world markets. While the increase in crude oil prices could help to mitigate this issue, it is not clear if a large share of the population will benefit from it.

Currently, the cereal-import dependence ratio for Yemen is estimated at 97% (Figure 2), with Russia and Ukraine accounting for the largest share of wheat and wheat product imports (42%). It's important to note that MENA countries (the source of 12% of Yemen's imports) re-export flour and other processed wheat products made from Black Sea-sourced products. Thus Yemen's exposure to the current crisis extends beyond its direct imports from Russia and Ukraine.

Figure 1: Evolution of Yemen's exports and imports, current US dollars, millions

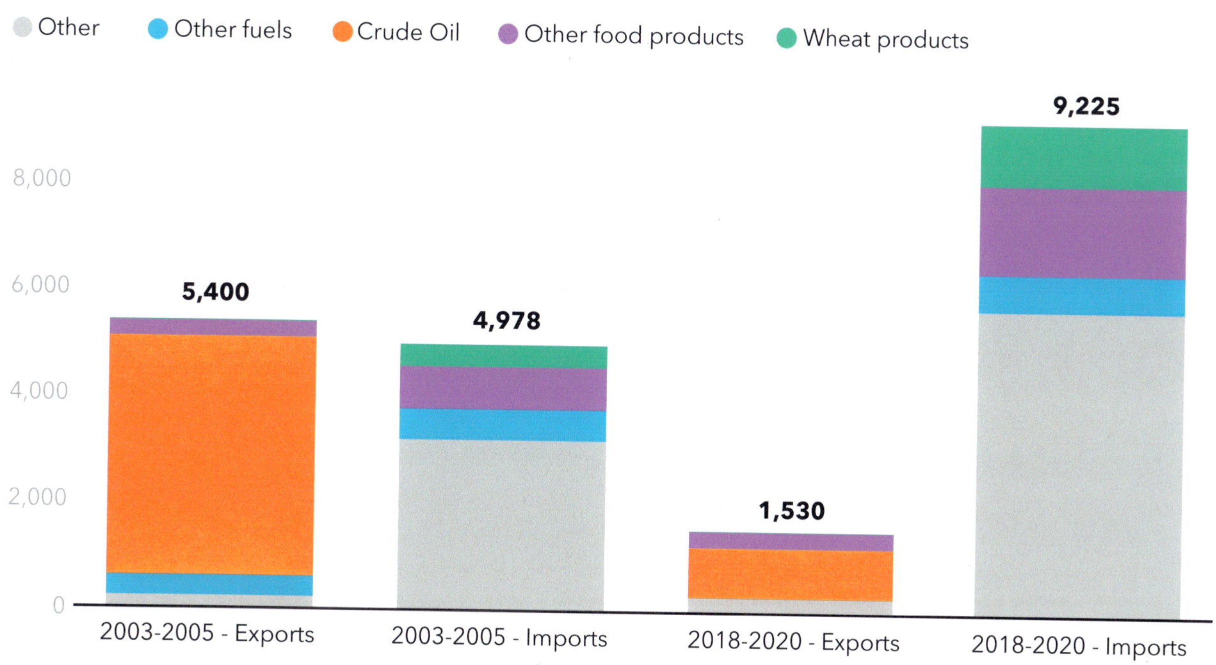

Source: COMTRADE

Figure 2: Daily per capita calorie consumption in Yemen, by origin and product, average 2018-2020

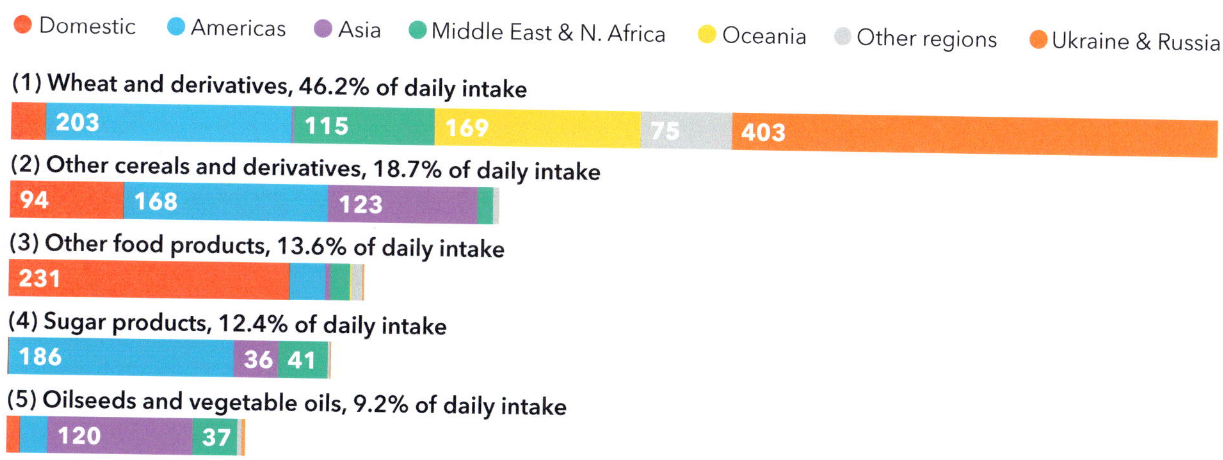

Source: IFPRI based on FAOSTAT & COMTRADE

Yemen's internal strife

What are the likely impacts on food security as the external shocks from rising prices and supply disruptions of wheat and fuel affect Yemen's already-tenuous internal circumstances?

In spite of the civil war and challenges of transportation, food markets currently remain generally functional; in the case of wheat, thanks to a robust network of small traders. Thus access to food for Yemeni households is primarily a challenge of affordability rather than a lack of availability. However, deteriorating conditions, including destruction of infrastructure since the start of the civil war in 2015, have systematically impoverished many Yemenis, prices have risen, and food insecurity has grown (Figure 3).

Figure 3: Current acute food insecurity situation, January-May 2022

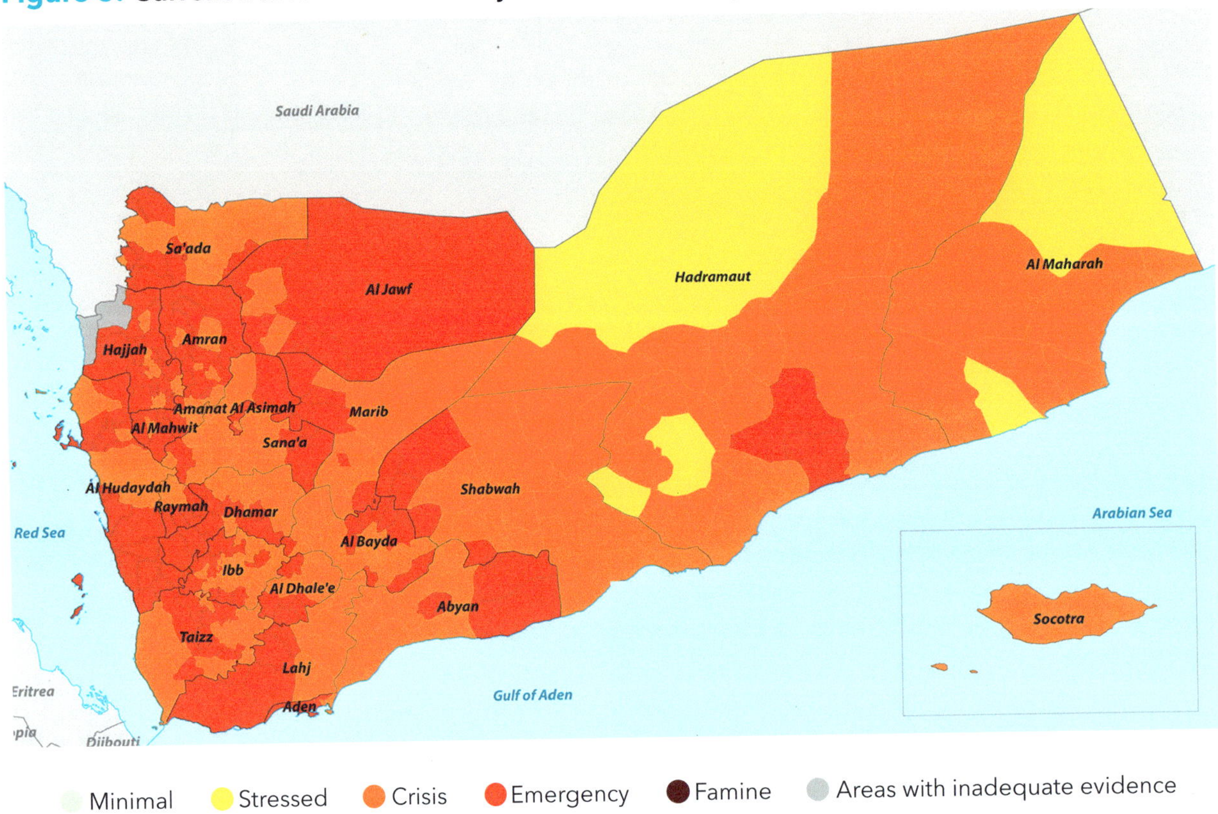

The conflict has divided the country into zones of control that affect flows of goods – particularly food and fuel. High fuel prices are already constraining Yemen's household mobility, access to water supplies, internal trade, and agricultural irrigation; as fuel prices climb higher in the wake of the Russia-Ukraine conflict, these problems will almost certainly intensify and with them, food insecurity. Yemen produces a modest amount of crude oil but lacks refining capacity, so it relies on imports of gasoline and diesel. The fuel market, meanwhile, is highly controlled by the government and there are severe fuel shortages with differences in regions controlled by the internationally-recognized government (IRG) and the de facto authority (DFA). Estimates indicate that local transport and distribution costs and taxes represent about three-quarters of the consumer cost of fuel in areas controlled by the DFA and two-thirds of the cost in areas controlled by the IRG.

Why wheat demand may increase despite rising prices

Meanwhile, rising wheat prices are a severe trap for Yemeni households that have limited alternatives for accessing sufficient calories given their limited income. In the most recent nationally representative household budget survey (2014, just prior to the civil war), 41% of the population had insufficient total calorie consumption, and poor households received 71% of total calories from staple grains, including wheat. But even as wheat prices increase and food expenditures rise, the primary response of Yemeni households is likely to be cutting out high-priced food goods and relying more heavily on the cheapest calorie sources – particularly imported wheat. This problem likely won't be solved by ramping up local production, which is very low compared to consumption demand; farmers also face the constraints of high fuel prices that make the use of diesel pumps for irrigation problematic and the lack of infrastructure oriented toward cereal production.

Role of the humanitarian response

Humanitarian assistance has been essential in buffering the impacts of rising food prices and other problems Yemeni households face. Approximately 12% of Yemen's cereal imports come in the form of humanitarian food aid. More than 50% of Yemeni households reported receiving some type of in-kind food assistance in 2020, according to phone survey data. The World Food Programme's assistance to Yemen alone aims to reach 13 million people – 43% of the population.

But the Russia-Ukraine war also threatens the continued flow of assistance, as it includes processed products such as fortified wheat flour and ready-to-use supplementary food (RUSF) that are often sourced from other MENA countries (Egypt, UAE, Türkiye) – which in turn import wheat from Ukraine and Russia (Figure 4). International funding for maintaining current levels of support is also uncertain as the Yemen crisis drags on and competing global priorities such as domestic COVID-19 responses and the refugee crisis in Ukraine have emerged. At the recent high-level pledging event on March 17, 2022, the donors committed approximately 30% less than in 2021 – in spite of the increased appeal based on higher projected needs as food prices increase and the emerging threat of famine.

Figure 4: Products imported into Yemen by the World Food Programme, share by value

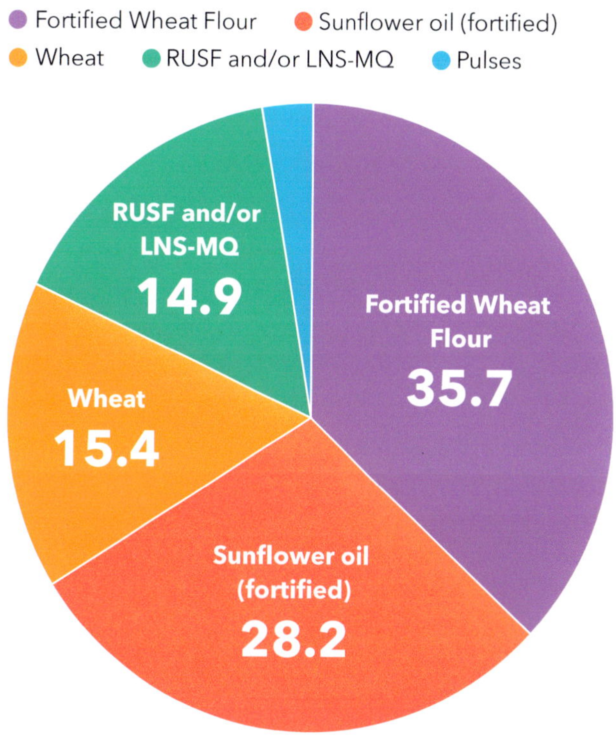

Source: IFPRI computations based on WFP procurement publicly available data

Potential responses to Yemen's predicament

Given the severity of these problems, what can the international community do to ease the growing pressure on Yemeni households from rising food and fuel prices?

One response is to increase the use of cash transfers and vouchers, which are more cost-efficient than in-kind food distribution. Reported overhead and administrative costs are 77% for in-kind distribution versus 11% to 17% for cash and vouchers. This approach also takes advantage of the robust domestic wheat trade, which has been surprisingly effective at adapting to the conflict setting. In addition, research by IFPRI and other organizations has shown that cash transfers allow households to increase consumption of non-staple food items and are effective at protecting children from malnutrition.

A challenge for this approach would be maintaining the value of cash transfers in the face of high inflation. Although the official exchange rate is stable at 250 YER to the dollar, the parallel exchange rate has reached approximately 1,200 YER to the dollar with high volatility in IRG areas of Yemen. Pegging transfer values to food basket prices or using hard currency as the transfer medium are alternatives employed in similar contexts in other countries.

Finally, responses to the current emergency should not neglect longer-term problems. Yemen's vulnerability to the commodities shock highlights the importance of a humanitarian–development nexus approach. Organizations should also focus on developing conflict-resilient approaches for local agricultural production that offer protection from expected shocks to fuel supplies and disruptions in long-distance transportation. Examples of such conflict-resilient agriculture projects in Yemen include drip irrigation and renovation of traditional spate irrigation and terrace systems, crops that are less reliant on irrigation and oriented to local markets rather than export, and solar-powered pumping systems.

Incorporating elements of resilience building and economic development into international support to Yemen is an important step. But ultimately, Yemeni people will only have the chance to prosper as "Arabia Felix" again if the main conflict parties exert more effort to end the civil war.

28 Addressing the food crisis in Yemen: The private sector's key role amid local conflict and global market disruptions from the Russia-Ukraine war

Sikandra Kurdi, Olivier Ecker, Joseph Glauber, and David Laborde

Originally published November 21, 2022.

The Yemen conflict, underway since early 2015, has led to an ongoing, unprecedented humanitarian emergency. Food needs far exceed current consumption levels, with 3.5 million pregnant or breastfeeding women and children under age five suffering from acute malnutrition and up to 19 million people affected by food insecurity in 2022.

Since February 2022, meanwhile, the Russia-Ukraine war has disrupted global supplies of grains and other key agricultural products and driven global food prices higher. Yemen depends heavily on grain imports to feed a population long teetering on the edge of famine. Maintaining wheat flowing into the country and wheat products reaching consumers through private sector importers, processors, and distributors is a critical puzzle piece for managing food security.

Both the conflict and the global market disruptions have placed extraordinary stresses on the private sector in Yemen. At a time when humanitarian efforts to address the crisis have been cut back, food importers and other private sector actors – always key to the country's food economy – have become central to the urgent task of preventing the further growth of food insecurity. In this post, we examine the current conditions that Yemen's private sector faces as it confronts these challenges and suggest a potential way forward to address related problems.

Food imports in Yemen

In 2020, 84% of food calories consumed in Yemen were imported – the vast majority coming from grains, mostly wheat. Prior to the start of the conflict, Yemen imported about 3 million metric tons of wheat per year, primarily from Australia, Russia, Ukraine, and the United States, or about 100 kg per capita per year. Due to declines in local agricultural production, the reliance on grain imports has since grown, despite the additional hurdles it creates for importers. After an immediate drop around the start of the conflict, wheat imports reached 3.4 million tons in 2020.

Food is imported to Yemen exclusively by a small number of private sector companies and humanitarian agencies. Food imports by humanitarian agencies amounted to 0.9 million tons, mostly wheat, in 2020 and have stayed high because of the growing food crisis. The World Food Programme (WFP) – by far the largest food importer among the humanitarian agencies – aims to support nearly 13 million of the most vulnerable people (almost half of the population) with emergency food assistance in 2022. WFP distributed approximately 26% of Yemen's total wheat imports in 2020.

Yet the costs of WFP's food distribution scheme (currently projected at US$809 million for the next six months) have proven difficult to sustain indefinitely. Due to funding constraints and global grain price increases, WFP was forced to reduce the amount of food distributed to 60% of the standard food basket portions for more than half of beneficiaries in January 2022, and again in June to 45–55% of the standard portions for all beneficiaries.

Even the enormous scale of food distribution leaves the large majority of Yemen's wheat imports supplied by the private sector. With assistance being cut back, monitoring the capability of the private sector to supply Yemeni food markets is an urgent concern.

Impacts of the Russia-Ukraine war and India's export ban

In 2021, Yemen imported 44% of its wheat from Russia (24%) and Ukraine (20%). Russia's invasion of Ukraine disrupted those supplies and forced a scramble by Yemen's importers to find alternatives, with mixed results so far.

Overall, for the first eight months of 2022, wheat imports from three of the usual top four export countries to Yemen (Australia, Ukraine, and the United States) together were down by 12% compared to the same period in 2021, largely due to reduced exports from Ukraine (Figure 1). Russia – the fourth top export country until 2021 – stopped reporting trade data in February 2022, so it is unclear whether any wheat from Russia was exported to Yemen since the start of the Russia-Ukraine war.

Despite those shortfalls, however, total export quantities to Yemen during the first eight months in 2021 and 2022 are similar, with the losses in 2022 largely offset by increased exports from India and the European Union.

Figure 1: Wheat exports to Yemen, January to August, metric tons

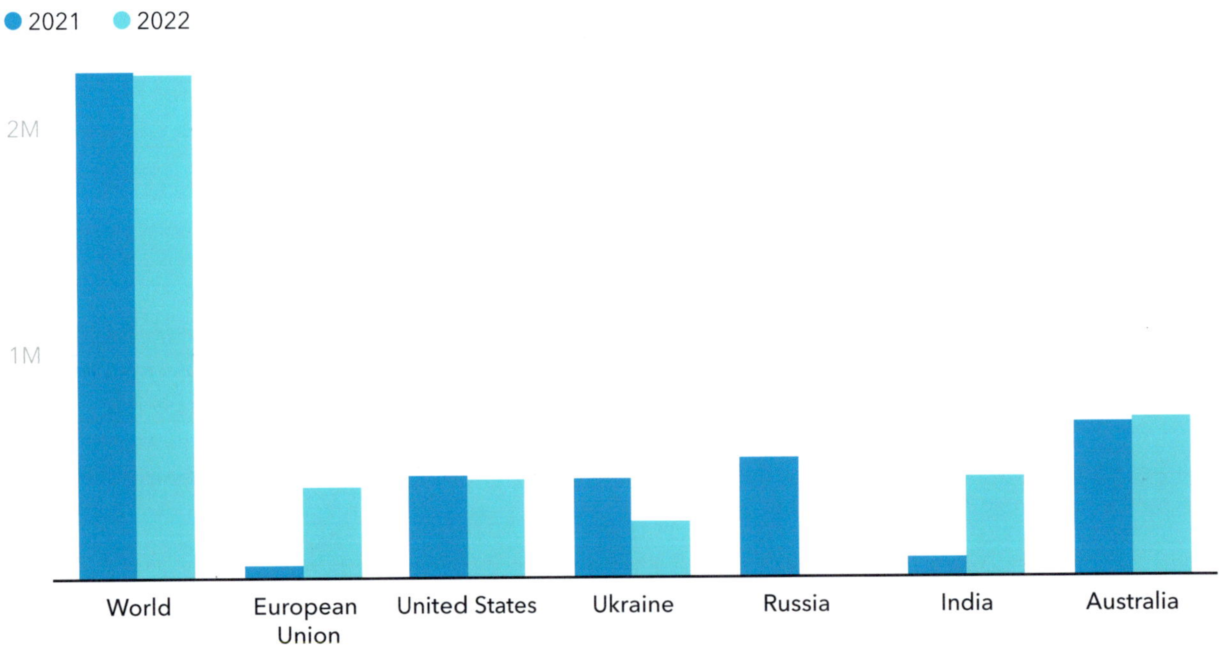

Note: World totals for 2022 do not include Russia. Russia suspended publication of trade statistics in April 2022.
Source: TDM

Thus, India's surprise wheat export ban in May 2022 hit Yemen particularly hard. India had become a crucial low-cost supplier contributing almost one-third of wheat imports to Yemen in the period from the start of the Ukraine conflict to the export ban.

While the Black Sea Grain Initiative has partially opened shipping from Ukrainian and Russian ports, only two shipments of wheat from Ukraine to Yemen have occurred so far, totaling 67,500 tons, with two further shipments in line for inspection. Shortly before expiration of the Black Sea grain export deal, the agreement was extended for three months on November 17. This extension provides crucial breathing room for getting further shipments of wheat to Yemen.

The price of the Yemen conflict

In the current global scramble to find alternative trade arrangements, the Yemen conflict has put food importers there at several disadvantages. The expiration of the United Nations-backed nationwide truce on October 2, 2022, which had held since April, is expected to weaken buyers' negotiating position regarding internationally traded commodities and create fresh logistical challenges for discharging grain shipments and distributing grain products in Yemen.

A greater concern is higher costs – a direct consequence of the conflict – being passed on to consumers. Since the conflict began, the price of insurance for commercial shipments has increased 16-fold. In addition, there are costly delays in delivering grain products due to the conflict-era system of clearances and inspections at the port of entry and along supply routes. Add to that the costs resulting from input supply constraints and local procedures. Prices of fuel – for operating port terminal generators, flour mills, and trucks, for example – are very high, and fuel shortages are common.

The de facto separation of the country into northern and southern parts requires supplying the North by imports through the seaports of Al-Hodeidah and As-Salif on the Red Sea, and the South by imports through the seaport of Aden on the Gulf of Aden. Distribution throughout Yemen's rugged countryside also raises costs, with higher prices seen in distant markets farther away from the main seaports and in remote areas with more complicated security situations.

Food importing companies in both North and South also face the complications in procuring foreign exchange to purchase in the international market. Each country part has separate monetary policies regarding the national currency – the Yemeni riyal (YER) – with more recent banknotes printed in the South not accepted in the North. The official exchange rate is overvalued, and importers must compete in an auction system to access foreign exchange at the Central Bank. In the South, the currency has been steadily depreciating overall while jumping unpredictably as government deficits climb and reserves dwindle amid occasional bailouts from Saudi Arabia (Figure 2).

The private sector as a whole in both North and South Yemen also faces obstacles in dealing with the international financial system, partly because the two parts are not recognized internationally as independent territories, and less favorable financing conditions are applied across the board. A particular challenge for importers is maintaining a much-larger-than-usual reserve of working capital (in US dollars) for financing shipments, which often require half of the import value paid upfront.

A related challenge is access to credit. Ansar Allah – the de facto authorities in North Yemen (also known as the Houthis) – and their supporters are targeted by various international financial sanctions strongly supported by the United States. The group was briefly designated as a terrorist organization by the Trump administration in January 2021, a move revoked the following month by incoming President Biden. While the current sanctions

Figure 2: Trend of parallel exchange rates in North and South Yemen, Yemeni riyal per US dollar

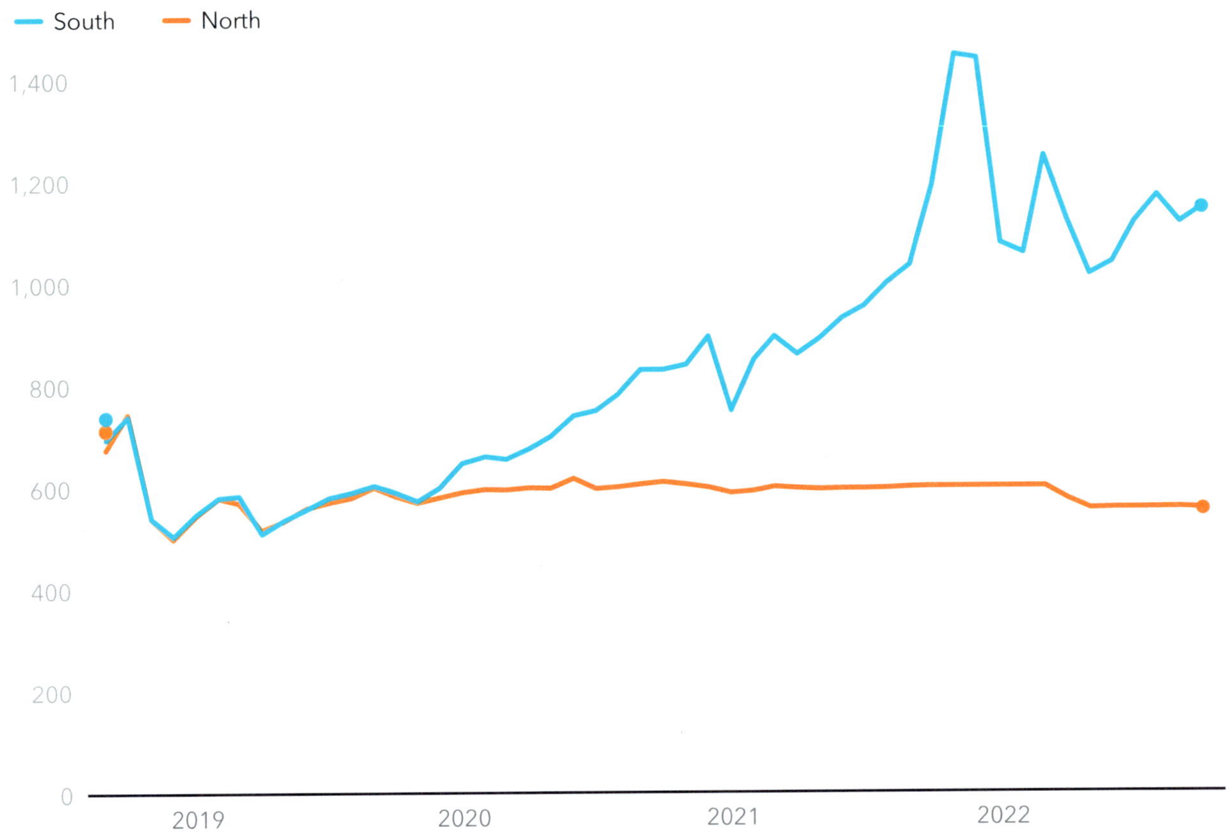

Source: Yemen Joint Market Monitoring Initiative, Situation Overview, September 2022

are in theory more narrowly targeted at Ansar Allah leadership and accomplices, in practice international banks prefer to avoid engaging in financial transactions involving Yemen as a whole.

All of these complications put upward pressure on food prices. Since the conflict began, local retail prices for wheat flour have trended at around twice the international wheat price. In the North, authorities have attempted to control price spikes, publishing a list with enforced price ranges for essential commodities at regular intervals. While this policy may benefit consumers in the short term by limiting price fluctuations, to the degree that prices are not updated sufficiently to reflect increases in costs, these price ceilings could contribute to food shortages in lieu of higher retail prices.

Maintaining food supply chains and consumer markets

While a few large companies dominate the food import and processing sector, a network of small traders and shopkeepers handles the distribution and retail sales of grain products in Yemen. These businesses play a pivotal role in Yemen's local economy, creating much-needed service sector jobs while also functioning as lenders for everyday essentials, including food. A recent survey of market key informants found that 64% of food vendors sell on credit to trusted customers in their communities, and most rural households heavily depend on these credits, especially for smoothing consumption of food purchases during periods of cash shortages. Approximately half of poor households' weekly food purchases are made using credit rather than cash. The demand for credit extends up the value chain. Wholesalers and other food distributors often buy on credit from importers and processors and in turn extend credit to these local retailers. While the provision of credit is vital for food security at the household level, it also implies challenges for liquidity in a context where access to financing is limited.

A way forward

One prong of any approach to address Yemen's current food crisis must be providing adequate funding for the humanitarian response to avert famine and reduce hunger. But it is also clear that, with humanitarian aid under pressure, making room for the private sector to operate effectively is key for food security in the country. The Hayel Saeed Anam (HSA) Group – Yemen's leading grain importer and processor of nearly all imported wheat – has called on international organizations and governments to provide emergency support for importers to avert a potential famine. Proposals include giving Yemen's wheat importers priority access to wheat supplies, providing them with special import financing conditions, and guaranteeing payments for shipments for a standardized 60-day term.

As long as the private sector can ensure that food remains available in local consumer markets, the international community can scale up cash assistance programs – these require less infrastructure than food distribution or food voucher programs and have been shown to be effective also under conflict conditions in Yemen. From the governments in North and South Yemen, urgent action in macroeconomic policy is needed to temper inflation and restore confidence in the Yemeni currency.

Yemen faces the continuing risk of soaring hunger and malnutrition amid both conflict and global economic uncertainty. Direct food aid and other humanitarian assistance is one prong of the response, but most wheat still reaches households through private importers and distribution networks. The international community can also take actions such as providing poor households with aid in the form of cash and take concrete actions to buttress the crucial role of the private sector.

29 West Africa faces mixed food security impacts from the Russia-Ukraine conflict

Antoine Bouët, David Laborde, and Fousseini Traoré

Originally published April 1, 2022.

Food security in West Africa has been deteriorating since 2015: The proportion of the population affected by undernutrition rose from 11.5% in 2015 to 18.7% in 2020, a total of 75.2 million people. Now, like the rest of the world, the region faces rapidly growing impacts from Russia's war in Ukraine, including spiking food prices and disruptions in markets for cereals and other commodities, including fertilizers and fuels. What are some of the war's likely effects on West Africa?

Overall, the greatest risk is from rising global prices. The region imports relatively little food or fertilizer (with some exceptions) from Russia and Ukraine, insulating it from many of the current market disruptions, and has capacity to substitute some lost fertilizer imports.

Commodity markets in turmoil

In February 2022, before the conflict began, the Food and Agriculture Organization's food price index had already reached an all-time high of 140.7 due to rising prices for vegetable oils, dairy products, and grains. Now, prices are spiking still higher as Russia and Ukraine account for a major share of the world's cereal exports, particularly wheat, and the conflict disrupts markets for fuels and other commodities. International fertilizer prices, for instance, have risen sharply: the price of urea, for example, doubled in February 2022 compared to its 2015–2019 average.

The good news for West Africa is that it trades little with Russia and Ukraine, except for Benin's exports to Ukraine (4.7% of its total exports) and Benin and Senegal's imports from Russia (4% and 5% of total imports, respectively). In terms of food security, the caloric contribution of these trade relationships is relatively small (Figure 1). Most of the calories consumed in the region come from local production, Europe, and Asia. Ukraine and Russia are significant suppliers of calories to West Africans only through trade in wheat (1.7% of total calories consumed). Ukraine's contribution is marginal, Russia's is more important.

The greatest challenge for West Africa is likely to be indirect effects of higher world prices (Figure 2). Before the Russia-Ukraine conflict, West Africa was seeing an economic recovery from impacts of the COVID-19 pandemic; the region's GDP per capita at constant prices had previously fallen by 1% in 2020, according to the International Monetary Fund (IMF). Now, rising prices threaten that recovery.

Indeed, West Africa is an importer of cereals, sugar, vegetable oils, meat, and animals, and an exporter of cocoa, cotton, groundnuts, fish, and fruits. Figure 2 shows increases in the price of palm oil, sugar, wheat, and chicken meat from January 2020 to February 2022, while the prices of bananas, cocoa, and groundnuts are stable. The only good news on the agricultural side is that the price of rice (imported) is stable (though international price

Figure 1: Daily per capita calorie consumption in West Africa, by origin and product, average 2018-2020

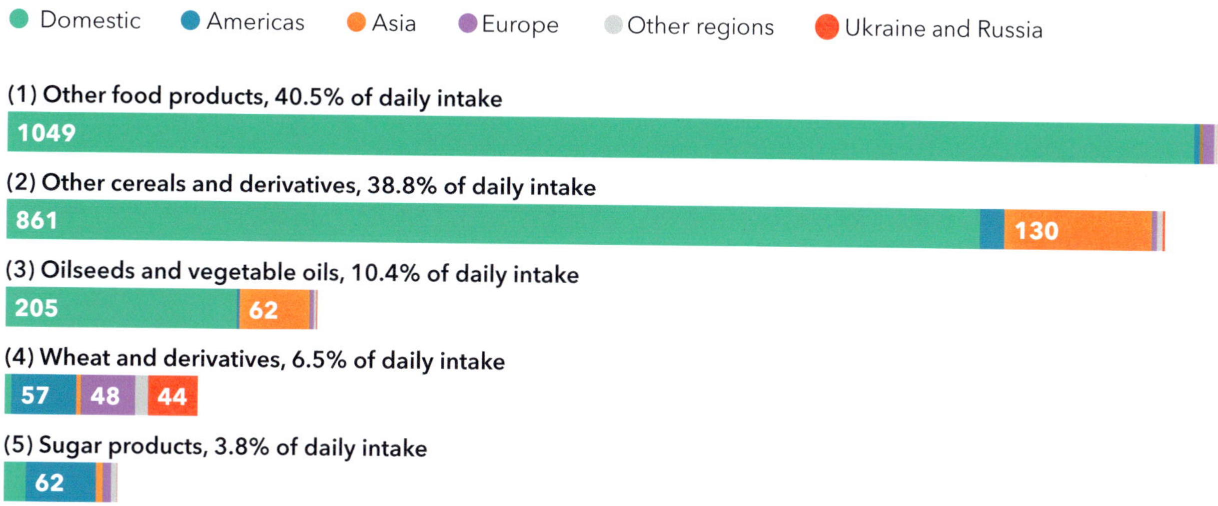

Source: IFPRI based on FAOSTAT & COMTRADE

Figure 2: Evolution of key commodity prices for West Africa

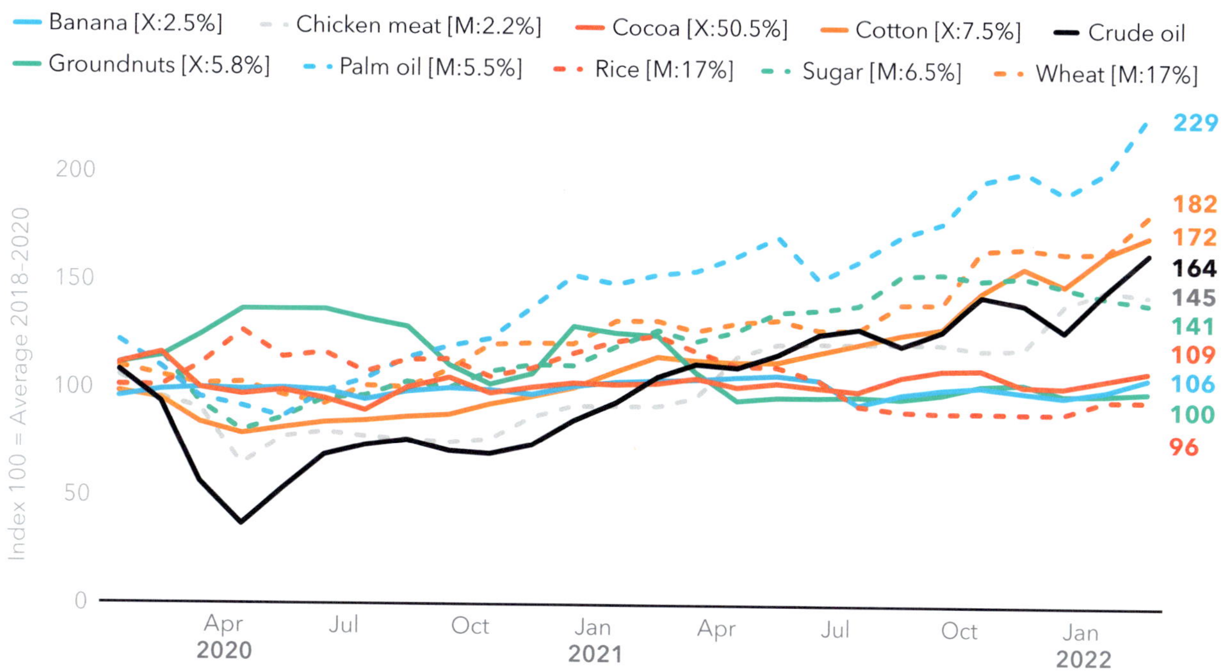

Note: Dotted lines indicate key imports. Solid lines indicate key exports. Trade status indicated in squared brackets with the percentage in the value of agri-food exports (X), or imports (M), of ECOWAS.

Source: World Bank, Bloomberg

volatility has recently increased), and the price of cotton (exported) has increased.

Meanwhile, outside of agriculture, commodity price developments are positive for the region, as it is also an exporter of certain raw materials: gold, tin, oil (which alone accounts for more than half of the region's export earnings), copper, natural gas, uranium, zinc, and aluminum.

Of course, the impacts on local populations of these various market shifts will depend to some degree on government policies. Increases in agricultural and food prices can only be compensated for by increases in commodity prices if there are redistribution mechanisms that allow for this compensation: For example, by taking revenues from mining rents to finance food subsidies or other forms of support. Price increases also pose a particular problem for urban households, which are more dependent on imports, such as wheat.

Fertilizers matter

The cessation or reduction of fertilizer imports from Russia – the second largest supplier to the Economic Community of West African States (ECOWAS), with 12% of the market – is also a serious risk for West African farmers. But it is also an opportunity for regional suppliers.

First and foremost, Nigeria has the capacity to make up for the missing imports. The country has seen heavy investment in the fertilizer sector through Indorama Nigeria and the Dangote Group, taking advantage of its sizable natural gas reserves. In addition, Morocco, which supplies one-third of the ECOWAS fertilizer market,

Figure 3: Share of the Russian Federation and Belarus in potassium fertilizer imports, by country, West Africa

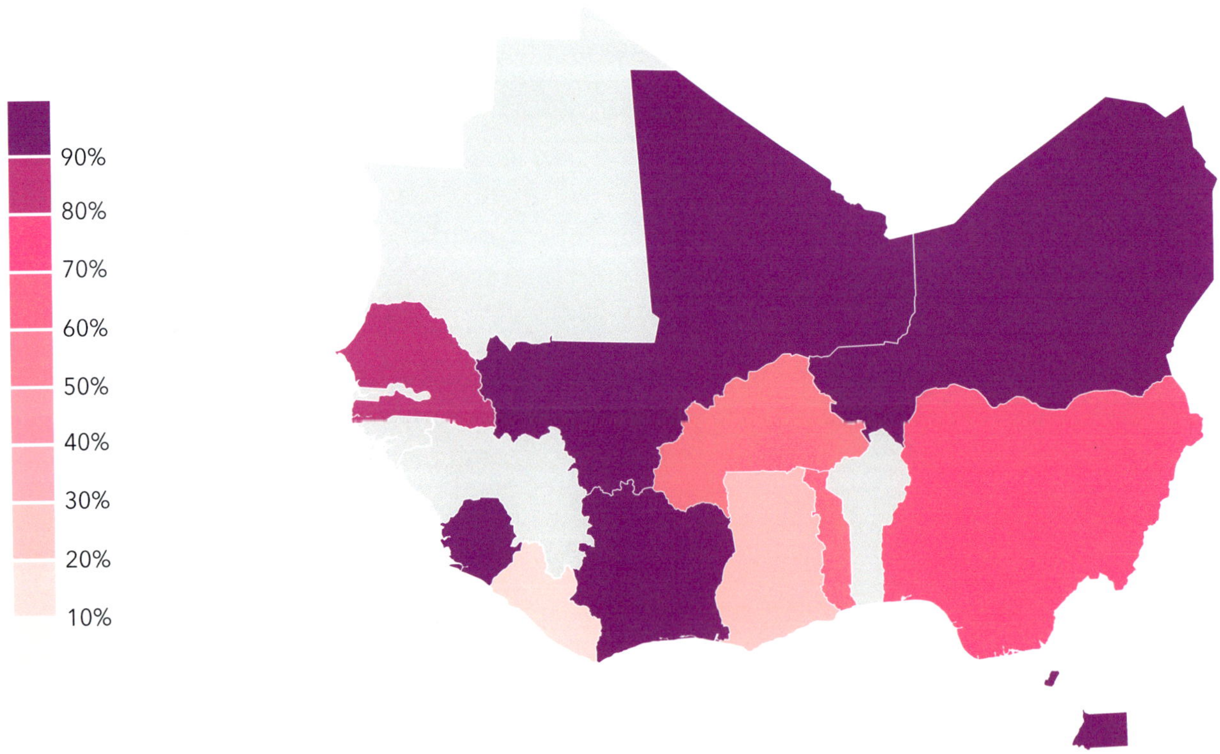

Source: COMTRADE

recently invested US$1 billion in Nigeria to build two phosphate plants (capacity of 1 million metric tons each). If we take the seven major West African countries, excluding Nigeria, for which data are available (Burkina Faso, Côte d'Ivoire, Ghana, Mali, Niger, Senegal, Togo), their fertilizer imports from Russia total only about 0.24 million tons. Nigeria could therefore remedy this imbalance: Dangote has a capacity of 3 million tons and Indorama 1.4 million, while Nigerian consumption is only 1.5 million tons.

However, difficulties arise for other types of fertilizers. West Africa imports 82% of its potassium consumption, and for Côte d'Ivoire, Mali, Niger, Senegal, and Sierra Leone, the vast majority of imports – 80% – come from Belarus and Russia (Figure 3).

Finally, farmers' access to fertilizer varies greatly from country to country. In some countries, only market mechanisms play a role, and price increases should be felt entirely by farmers. In others, the cost is largely borne by public budgets through state subsidy programs (Nigeria, Ghana, Togo).

An initial context of food inflation and political conflict

Economic impacts of the Russia-Ukraine war are also intersecting those surrounding the 2021 coup in Mali. In response to the coup, ECOWAS has twice imposed an embargo on the country, effectively closing its borders to all member states, except for trade in basic necessities. Malian authorities have twice taken retaliatory measures. Some Malian sectors, such as livestock and meat, had already prohibited their members from exporting within ECOWAS. As Mali is one of the region's main suppliers of livestock, the suspension of Malian exports has created market tensions – particularly in the meat sector – in some countries. Since the latest retaliation, the price of meat has risen by 25–30% in Dakar, Senegal.

The economic fallout from regional tensions in West Africa and the Russia-Ukraine situation illustrates the dangerous links between diplomatic sanctions, trade, and food security. These situations continue to play out. Mali is contesting the legality of the embargo under EU law and the United Nations Convention on Trade in Landlocked Countries. The current situation presents a challenge for ECOWAS, which should refrain from taking measures that could worsen regional food security, already greatly aggravated by the COVID-19 crisis. Intra-ECOWAS trade of agricultural and food products should be facilitated and not restricted by measures like export restrictions or embargos; regarding the rest of the world, a reduction of the common external tariff, already low on food products, should be considered, as well as redistributive measures to support poor households.

30 The Russia-Ukraine conflict is likely to compound Sudan's existing food security problems

Clemens Breisinger, Oliver Kirui, Paul Dorosh, Joseph Glauber, and David Laborde

Originally published April 6, 2022.

Russia's invasion of Ukraine has disrupted agricultural production and trade from one of the world's major food exporting regions. The war threatens to drive rising food prices still higher and create scarcity, especially for regions most dependent on wheat and other exports from Russia and Ukraine – particularly the Middle East and North Africa.

Sudan faces a uniquely difficult set of circumstances as these disruptions loom. As with other countries in the region, wheat is a key food item for Sudan, second only to sorghum as a source of calories and accounting for 530 calories/person/day – a fifth of the estimated 2,576 total calories consumed daily. Only about 15% of the wheat consumed is grown in Sudan – a share that might shrink due to rising fertilizer and energy prices; the rest is imported, with a majority sourced from Russia and Ukraine in recent years (Figure 1). Adding to these vulnerabilities, prices for wheat and fuel were already spiking before the war began, compounding the risk of rising food insecurity.

Figure 1: Sudan's wheat imports by origin, million metric tons

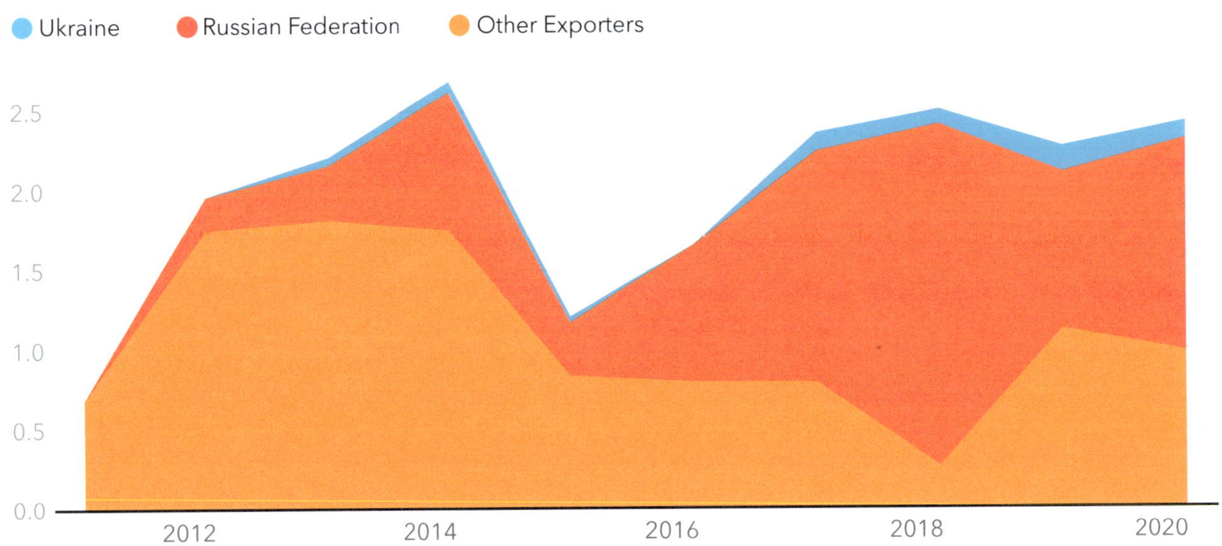

Source: COMTRADE

What led to this situation, how serious is it, and what can be done to head off future crises?

Wheat and bread play a central role for food security and political stability in Sudan. Demand for wheat has grown rapidly in the last 15–20 years, especially in urban areas where 35% of the population lives, driven mainly by population growth and changing consumer preferences for bread and other wheat products. Wheat is thus politically important, and large-scale protests have followed changes in the subsidized price of baladi bread (traditional flatbread) in recent years. For example, a sharp increase in the sales price of baladi bread in December 2018 led to massive street protests.

About 85% of Sudan's wheat supply has come from imports over the past decade, costing the country about US$500 million a year (FEWS NET 2015; World Bank 2020). Domestic wheat production, which accounts for the remainder of supply, is concentrated in El Gezira State in the Gezira Scheme, one of the world's largest irrigation projects. The average national yield is low (about 2 metric tons per hectare), and increasing area for wheat cultivation may be possible but would have important trade-offs with other higher-value agricultural goods.

Wheat prices and volatility had increased sharply before the Russia-Ukraine conflict, the result of a series of domestic problems, including high inflation and political instability, with severe adverse consequences for the Sudanese economy. Wheat prices began rising in 2019 and have recently spiked (Figure 2).

In particular, the most recent surge in wheat prices in 2021 is due largely to overall domestic inflation, foreign exchange shortages that limited wheat imports, a rapid depreciation of the exchange rate, and continued low productivity of domestic production. In addition, the removal of fuel subsidies in June 2021 contributed to increased production costs for farmers and bakeries for non-subsidized inputs such as water, yeast, cooking gas, labor, and oil. The inflationary pressures were compounded further by a cut in funding from international donors following the military coup in October 2021. Many bakeries went out of business as production costs increased more than the official sales price of subsidized flatbread.

Figure 2: Domestic wheat and bread prices, Sudanese pounds per kilo

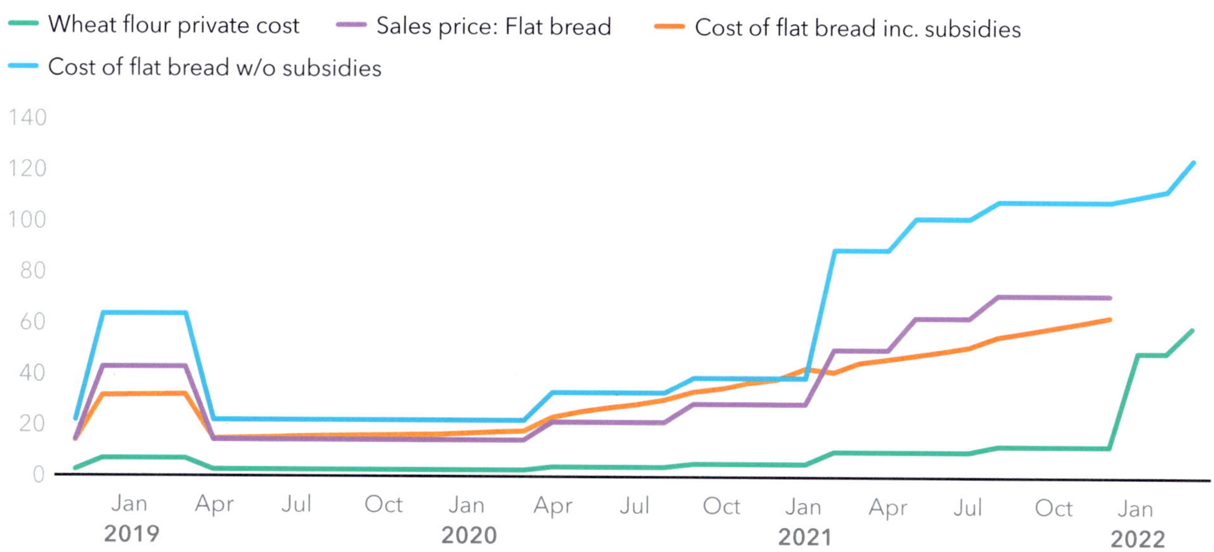

Note: Wheat flour private cost includes the subsidy (prior to January 2022).

These problems intensified with the arrival of 2022. On January 1, Sudan's government abandoned all forms of subsidies on wheat (grain, flour, and bread), forcing milling companies to obtain grain in the higher-priced open market (Al-Nilin 2021; Abay 2021). Overall, between July 2021 and February 2022, the wholesale price of wheat in Khartoum rose by 112% (about 60% in real terms).

Then came Russia's invasion of Ukraine. The resulting disruption of wheat exports from those countries has pushed imported wheat prices still higher. Bread prices are expected to rise further as well, both because of higher wheat prices and increased production costs due to higher oil and gas prices.

For example, the price of gas used as cooking fuel for most bakeries has recently jumped 56%; the price of a jerry can of oil has jumped 67%. Higher wheat and petroleum prices also add to pressure on foreign exchange reserves, contributing to a recent government decision to sell gold to fund additional food imports in advance of the upcoming Ramadan months, when household food consumption typically increases sharply. (Authorities have been forced to increase withdrawal of proceeds from the gold trade to pay for imports of strategic commodities.)

To better understand these market dynamics, we ran simulations using a partial equilibrium model. These suggest that the large increase in nominal and real wheat market prices between August 2021 and February 2022 is consistent with a 24% decline in wheat imports and a decline in total wheat consumption (including wheat consumed as bread) of 15%. The recent world price increase associated with the Ukraine war can be expected to lead to an additional increase in market prices of imports and domestic wheat of about 20%, and a further decline in wheat imports and consumer demand for wheat products of 9 and 5 percentage points, respectively (Figure 3).

Urban poor households are most affected, as their wheat consumption drops by 16% to 19% between July 2021 and February 2022 because of domestic policy changes, and then by another 5% in March 2022 (Figure 4). Percentage declines in wheat consumption are similar, but slightly less severe for other household groups. The welfare of rural poor households, which constitute the majority of the population, is least affected by these wheat market shocks given their relatively low per capita consumption of wheat products in 2022 (only about 12 kg/person).

Thus, the model indicates that disruptions in the world wheat market due to the Russia-Ukraine war have had a major impact on Sudan's wheat market—though the estimated effect is only about half the magnitude of the adverse shock of July 2021 to February 2022. Nonetheless, the combined effect of these wheat price shocks has serious consequences for Sudan's food economy and especially urban poor households.

Figure 3: Effects of import shocks on Sudan's wheat market

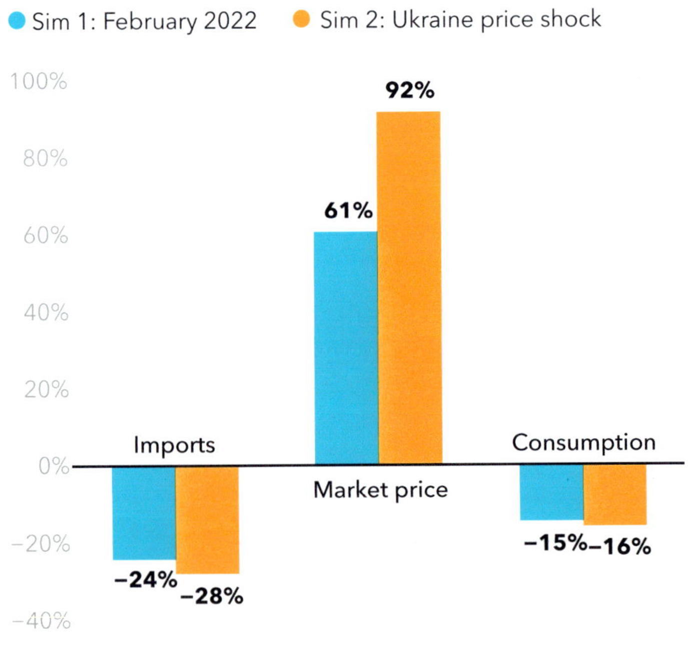

Source: Sudan wheat model simulations

Figure 4: **Effects of world wheat price shocks on household wheat consumption in Sudan, kilograms per person per year**

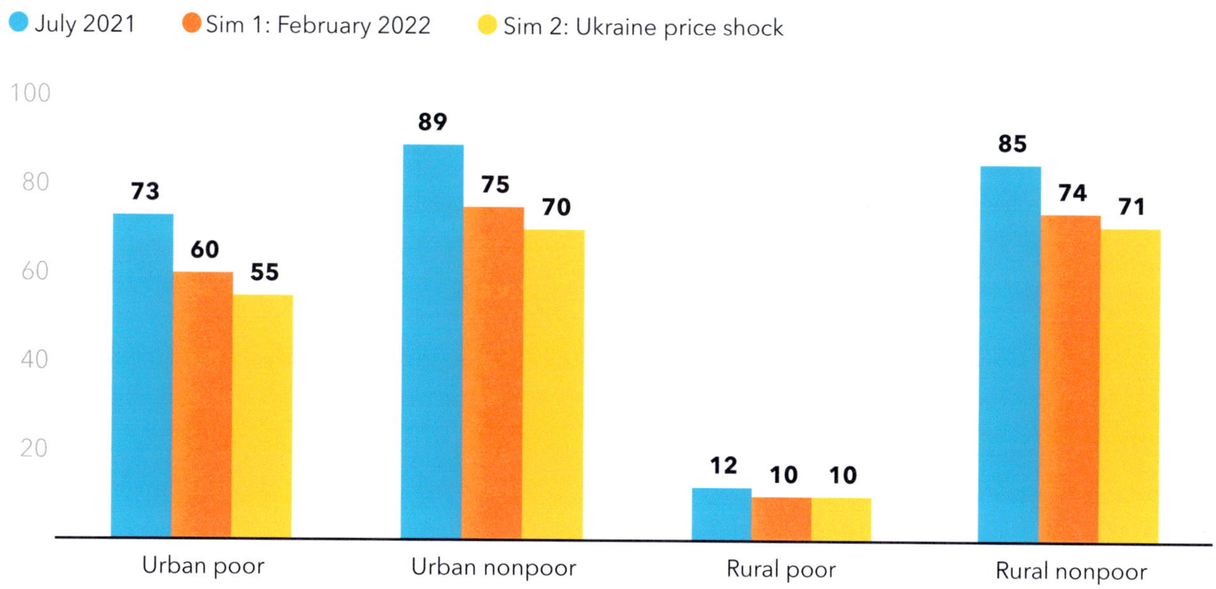

Source: Sudan wheat model simulations

In addition to addressing the immediate impacts of the current crisis, Sudan should look to building its resilience in order to reduce the impacts of the next one. Here are several policy measures to consider:

- Increased investments in roads and other market infrastructure to reduce transaction costs, raise producer prices and improve the functioning and efficiency of markets.
- Additional research and extension efforts to increase production of alternatives to wheat, including drought-tolerant sorghum and millet on non-irrigated land and high-value export crops on irrigated land. Such a shift in production has the potential to increase foreign exchange earnings and promote overall growth and employment.
- As bread subsidies are removed, a cash-transfer system targeting the poorest households could be considered. Such a program could build on the experiences from the Sudan Family Support Program (Samarat) introduced in 2021 but then halted, as well as lessons from successful cash-transfer program in other countries, including Egypt's Takaful and Karama programs. The size of the cash transfer would need to be carefully calibrated and perhaps adjusted frequently as wheat prices and inflation change.
- Food security data gaps need to be addressed and monitoring and analytical capacity should be increased. Accurate targeting of either cash or in-kind transfer programs would require up-to-date household data. Public sector capacity should also be enhanced to increase the effectiveness of standards agencies to enforce regulations for monitoring of wheat flour quality, ensure competition in wheat milling, and effectively build and manage a cash transfer program.

Putting some or all of these in place can help to build a more resilient wheat and agrifood system in Sudan, giving households the wherewithal to better withstand future economic shocks.

31 Russia's invasion of Ukraine threatens food security in Malawi: How can the country respond?

Joachim De Weerdt and Jan Duchoslav

Originally published April 27, 2022.

Food insecurity is endemic in Malawi, affecting up to 38% of the population every year in the run-up to the harvest in April. Although geographically distant, there are multiple channels through which Russia's invasion of Ukraine can make matters worse this year.

The conflict has disrupted global supplies of key cereals, vegetable oils, and fertilizers, pushing already-high prices higher, and putting particular pressure on low-income countries with vulnerable poor populations. In this post we discuss how rising wheat, maize, cooking oil, and fertilizer prices are likely to impact Malawi and how the government can respond.

Grain and vegetable oil markets in Malawi

Wheat

Rising wheat prices have been making global headlines and will affect many countries severely, but their direct impact on Malawi will be limited. Malawians rely on wheat for only 2.3% of their energy intake (Figure 1) and spend only 2.8% of their total food budget on wheat or wheat products like bread and pasta (Figure 2). Moreover, these products are mostly consumed by relatively richer and urban households, so the rising prices of bread will have only moderate direct effects on the country's food security. That said, imports from Russia and Ukraine made up nearly two-thirds of Malawi's wheat consumption in recent years and the country may face a shortage until alternative suppliers are contracted.

Maize

Ukraine and Russia jointly account for 17.4% of world trade in maize, and global maize prices are rising. Malawi is roughly self-sufficient in maize production in years with favorable rainfall, but is often a net importer. On average, the country imports less than 1% of its consumption, but at the extremes, net imports have risen to 10% of domestic consumption, as happened in 2016 when harvests were particularly poor; and net exports have gone up to 10% of domestic production after a good harvest, as in 2011.

This year, a late onset of the rains and floods in Malawi's southern region are likely to cause production shortfalls in parts of the country, although it is too early to tell how much this will impact total national production. One indication of rising maize prices is that the government's recently announced minimum farmgate price for maize is 47% higher than last year's. Maize imports and exports are heavily regulated, but the country is not at all immune to international price trends.

Figure 1: Share of maize, cooking oil, and wheat in Malawian diets, percent of calories

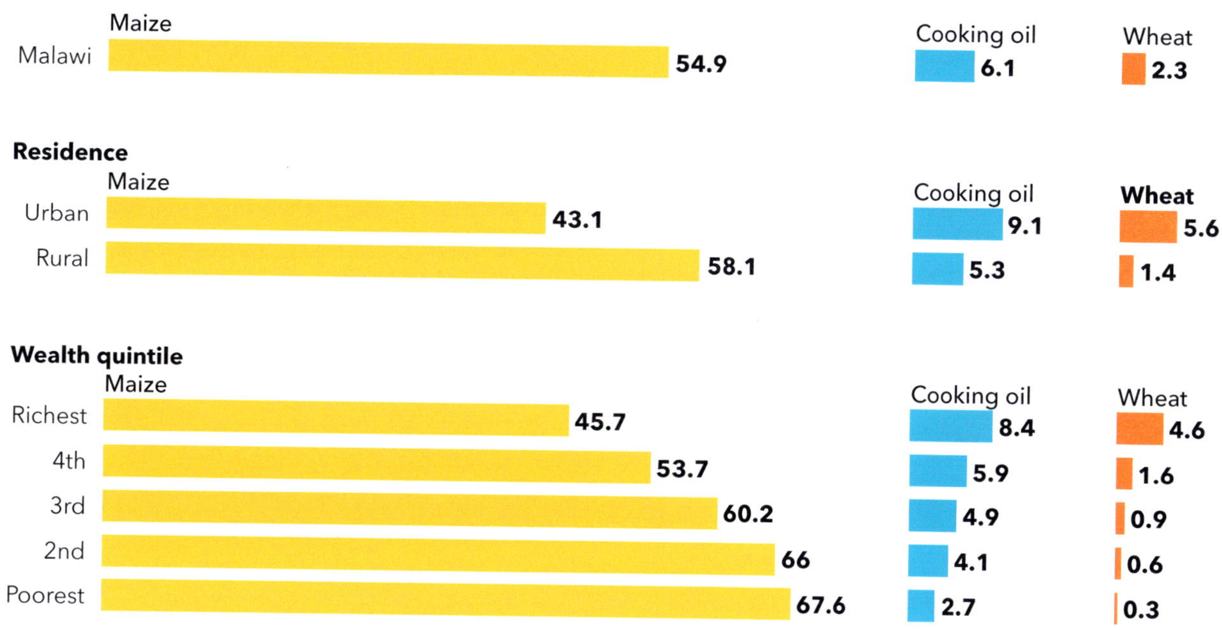

Source: Fifth Integrated Household Survey 2019-2020

Figure 2: Share of maize, cooking oil, and wheat in Malawian food purchases, percent of value

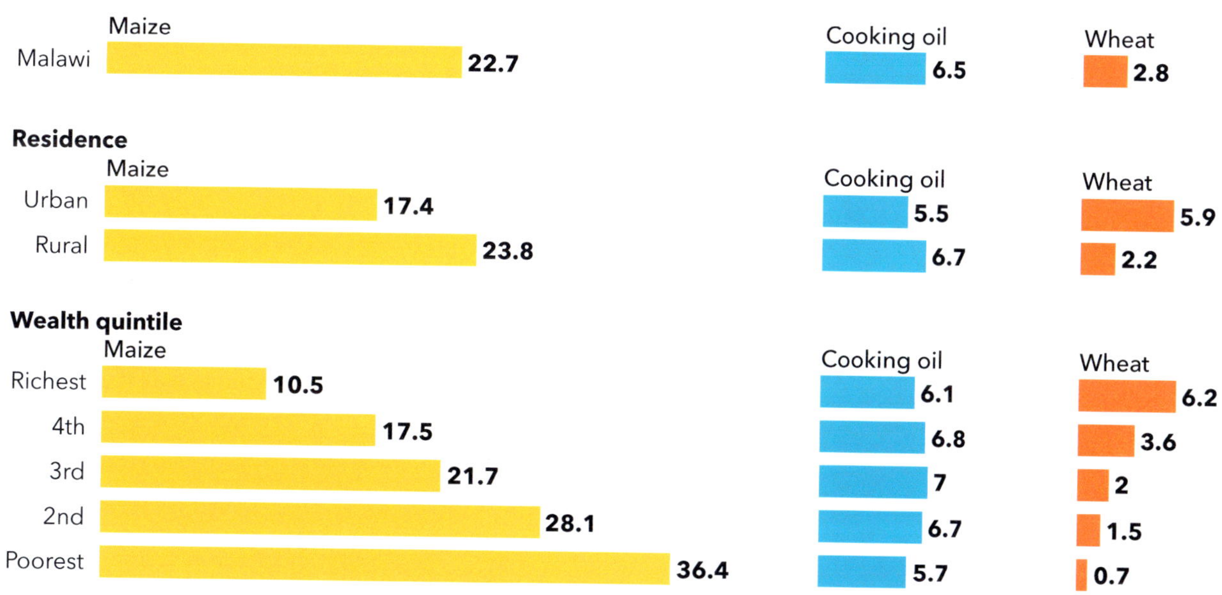

Source: Fifth Integrated Household Survey 2019-2020

Malawian households produce as well as consume maize. In contrast to wheat, maize prices will have a strong impact on the poor, who derive 68% of their calories from this source and for whom it makes up 36% of total purchased food.

Because high food prices benefit those who sell and harm those who buy, it is important to understand who produces and who consumes. While nearly all Malawian farmers grow maize, most of that is for own consumption. Figure 3 shows that richer households are less likely to grow maize, but when they do, they are more likely to produce more than needed at home. These households can potentially benefit from rising maize prices. Most of the poor, by contrast, grow some maize, but only 5% of the poor produce enough for their own needs.

Figure 3: Malawian households by maize production, percent

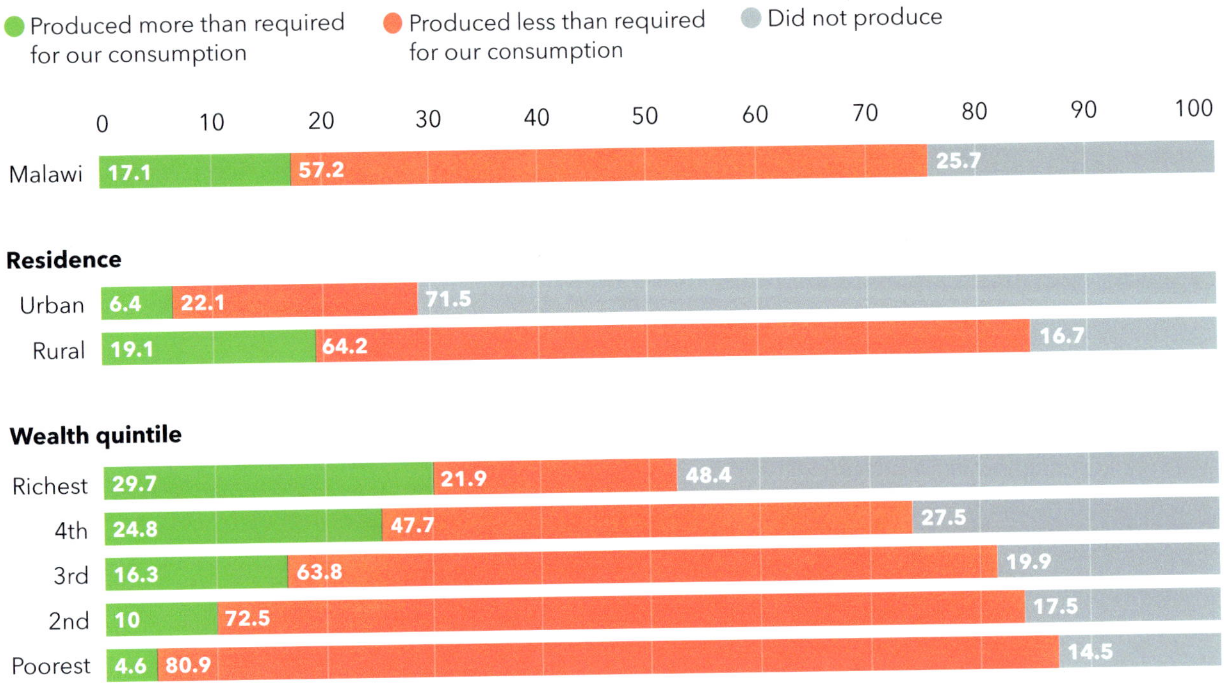

Source: Fifth Integrated Household Survey 2019-2020

While widespread maize production in Malawi will insulate households somewhat from a global maize price shock, most households will turn to the market at some point in time to purchase locally produced maize for consumption. Poor and urban households are most dependent on purchases of maize and will be hurt most by rising maize prices.

Cooking oil

Even before the war, global weather events had affected soybean and palm oil production, putting upward pressure on the price of vegetable oil; rising prices of cooking oil have been the subject of fierce political discussions in Malawi for months. With Russia and Ukraine jointly accounting for nearly three-quarters of the world's sunflower oil trade, further price increases should be expected.

Figures 1 and 2 offer a glimpse of how this translates to household budgets and calorie sources. Cooking oil makes up a larger share of calories for urban and richer households, but there is no discernable pattern when it comes to share of purchases – across all wealth quintiles and irrespective of urban or rural location, cooking oil makes up a stable 5.4% to 7.0% of total food purchases. This means that while high cooking oil prices are likely to be felt across all segments of the population, the impact will fall harder on the poor, who already consume little oil and therefore have less scope to further reduce consumption.

On the other hand, survey data show that 16% of crop-producing households in Malawi grew soybeans in 2019/20, a number which has likely increased since then because of favorable prices in the 2020/21 growing season. This group of households stands to benefit from the higher soybean prices, which are likely to rise further as cooking oil manufacturers seek substitutes for Ukrainian and Russian sunflower seed.

High fertilizer prices could put agricultural production at risk

Similar supply shocks will reduce the availability and drive up prices of chemical fertilizers. Before the war, Russia and Belarus jointly accounted for 15% of global trade in nitrogenous fertilizers and 33% of potash. When global fertilizer prices nearly doubled in the year to August 2021, it derailed the government's implementation of last year's Affordable Inputs Programme (AIP), which provides eligible farmers with fertilizer and hybrid seed at reduced prices.

Since then, prices have nearly doubled again, and are likely to continue rising. This does not bode well for the AIP, whose budget for 2022 has been reduced by a third in real terms compared to 2021. If the current prices persist and if the government wants to retain current subsidy levels, the number of beneficiaries will have to be reduced by roughly two-thirds. If the government wants to maintain the number of beneficiaries, their co-payment will have to drastically increase. Either way, less fertilizer is likely to be applied in 2022, meaning reduced agricultural production, putting still more upward pressure on local and global food prices.

Policy options

How can Malawian policymakers help steer the country through this crisis unscathed, and prepare for future crises? We lay out a number of complementary policies below.

First, rising food prices are compounded by seasonal fluctuations in maize prices. The government cannot do much to alter global market prices, making it more important than ever to stabilize seasonal fluctuations in maize prices.

Figure 4 shows how volatile maize prices have been over the past four years, rising between 25% and 105% toward the lean season, compared to the price in May right after harvest. Yet many Malawian farmers, cash-strapped after the lean season and lacking safe storage facilities, sell much of their produce shortly after harvest, only to buy it back at a loss later in the year. These fluctuations can be moderated via better management of the Strategic Grain Reserve (SGR).

In recent years, the SGR typically began its replenishment only after the start of the fiscal year in July, putting upward pressure on already rising maize prices. As of 2022/23, however, the start of the fiscal year has been shifted from July to April. This means that government agencies should have their budgets available in time to make the SGR procurement early enough to prop up prices when they are at their lowest, thus reducing the disparity between the buying and selling prices that Malawian farmers face. This would primarily benefit the poorest farmers, who are most likely to sell their harvest early and the most vulnerable to rising price levels in the country.

Figure 4: Seasonal maize price fluctuations 2018–2022, percent change since May

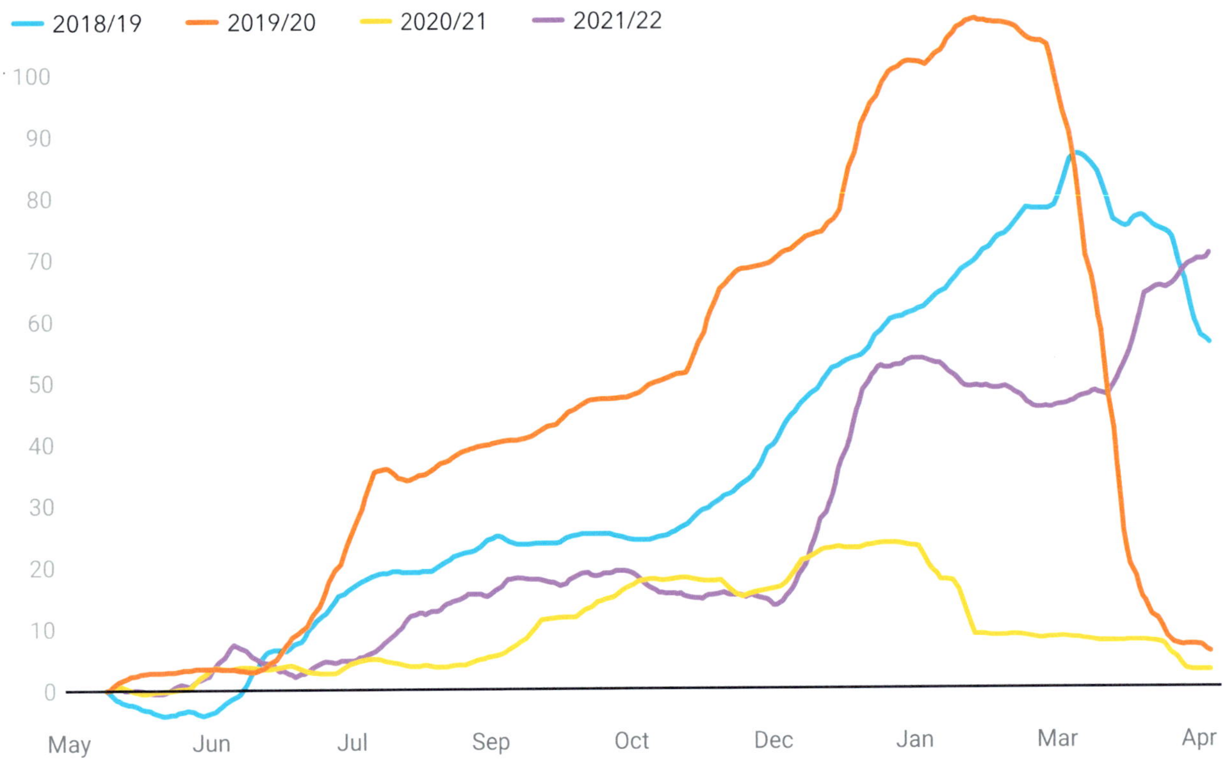

Source: Fifth Integrated Household Survey 2019-2020

Second, Malawi imports all its fertilizer, so unlike maize, there are few options to alter or smoothen the price of fertilizer. Policy responses should thus focus on increasing efficiency of fertilizer applications. Unfortunately, yield responses to fertilizer have been declining over recent years due to deteriorating soil health. Even before the Russia-Ukraine war, some farmers could not make profitable use of fertilizer, in the sense that the total cost of the fertilizer would be higher than the value of the additional yield resulting from its application. Further increases in the fertilizer prices will only exacerbate such inefficiencies.

Policy options to make more efficient use of fertilizer and fertilizer subsidies have been discussed at length by De Weerdt and Duchoslav (2022), Chadza and Duchoslav (2022), and Nyondo et al. (2022), who have called for increased investments in soil health, in agricultural research and development, and in agricultural extension.

Third, social safety nets should be agile and aware of who is affected by a particular shock. During the COVID-19 pandemic, the urban poor suffered disproportionately from lockdown restrictions. Rising maize prices are likely to impact the same group, which is completely dependent on food purchases that take a large chunk of their budget. Malawi's flagship social safety net, the Social Cash Transfer Programme (SCTP), is rigid, so ad hoc interventions such as the COVID-19 Urban Cash Intervention or the recurrent but ad hoc funded Lean Season Food Insecurity Response Plan have been deployed to respond to shocks. Making the SCTP sufficiently flexible both horizontally (that is, promptly increasing and decreasing the number of beneficiaries as needed) and vertically (that is, varying transfer sizes) would help to better protect the most vulnerable households from shocks like the current high food prices.

Finally, despite decades of policies that promote maize self-sufficiency at the household level, and despite three-quarters of Malawian households growing maize and over 50% of the agricultural budget allocated to subsidizing agricultural inputs primarily geared toward maize production, the vast majority of Malawian households depend on market purchases of maize.

Among the poorest 20% of the population, 86% grow maize, but only 5% are likely to produce sufficient quantities to satisfy their own consumption needs (Figure 3). The focus on subsistence is clearly not producing the desired results in terms of resilience to external shocks, while it is simultaneously crowding out investment in other, more productive activities.

We have already noted that making strategic use of the grain reserves will be especially important to bridge this volatile year, but in the longer run, diversifying agricultural production to more lucrative, commercially oriented crops, as well as diversifying beyond agriculture, are crucial goals. A necessary condition for such a shift is that farmers should be confident that they will have access to their staple food at predictable prices. At the national level, this can be achieved by ensuring that government actions in the maize market are predictable and rules-based, while at the international level, strengthening regional integration of maize markets can play a key role.

32 The Russia-Ukraine crisis presents threats to Nigeria's food security, but potential opportunities for the fertilizer and energy sectors

Bedru Balana, Kwaw Andam, Mulubrhan Amare, Dolapo Adeyanju, and David Laborde

Originally published June 9, 2022.

The current rise in global market prices for major food commodities almost mirrors that of the 2008 food crisis, presenting a worldwide threat to food security. The situation is particularly severe in Africa, where the COVID-19 pandemic and now the Russia-Ukraine crisis have exposed the vulnerability of food systems to major shocks, particularly in countries like Nigeria that rely heavily on imports of major staple foods such as rice and wheat.

With global food prices spiking and supplies of wheat, oils, and other items disrupted due to the Russia-Ukraine war, Nigeria faces a number of threats to its already precarious food security. Since over 50% of the foods consumed by Nigerian households come from purchased sources, food price inflation threatens to place many people in a worsening food insecurity situation. In particular, Nigeria's dependence on wheat imports may lead to high prices and supply problems. At the same time, however, Nigeria's capacity to produce other key items – in particular, fertilizer and natural gas – may allow it to take advantage of global market disruptions from the crisis.

In this post, we examine how wheat supply disruptions and spiking prices caused by the Russia-Ukraine conflict may exacerbate food insecurity in Nigeria, and also explore the country's potential opportunities in the emerging fertilizer sector and energy industries.

Nigeria's food security situation

Nigeria is particularly vulnerable to current spiking global food prices. With a population of nearly 217 million people (about 15% of Africa's population), Nigeria is the most populous country and the largest economy in Africa. However, like most countries in Africa south of the Sahara, Nigeria has high poverty rates, with 42.6% of people living below the poverty line, unemployment at 33%, and twin challenges of food insecurity and acute malnutrition (Ecker et al., 2021), with 35% of children under age five stunted, according to the Nigeria Demographic and Health Survey.

Nigeria is among the 10 countries with the highest number of people in food crisis, according to the 2022 Global Report on Food Crises: 12.94 million people were in acute food insecurity in October–December 2021 (the report's analysis covers 21 of the country's 36 states, and the Federal Capital Territory). As IFPRI studies on food security effects COVID-19 in Nigeria show, COVID-19-induced shocks have exacerbated the vulnerability and food insecurity of Nigerian households (Andam et al. 2020; Abay et al. 2021; Amare et al. 2021; Balana et al. 2020).

Nigeria's wheat imports (2016-2020)

Nigeria is a huge consumer and importer of wheat products. According to the Central Bank of Nigeria (CBN), wheat is the third-most-consumed grain in Nigeria after maize and rice, with domestic production accounting for only 1% of the 5 to 6 million metric tons of wheat consumed annually. As shown in Figure 1, Nigeria imported wheat worth over US$2.15 billion in 2020, a 40% rise from the previous year, and ranked as the world's fourth-largest importer of wheat after Egypt, China, and Türkiye in 2020, making the commodity the largest item on its import bill after petroleum products (petrol and diesel) and the highest imported food item in Nigeria (National Bureau of Statistics (NBS), 2021). Russia was the second-largest source of wheat imports in 2020 ($401 million) behind the United States ($518 million).

Figure 1: Nigeria's wheat imports, US dollars millions

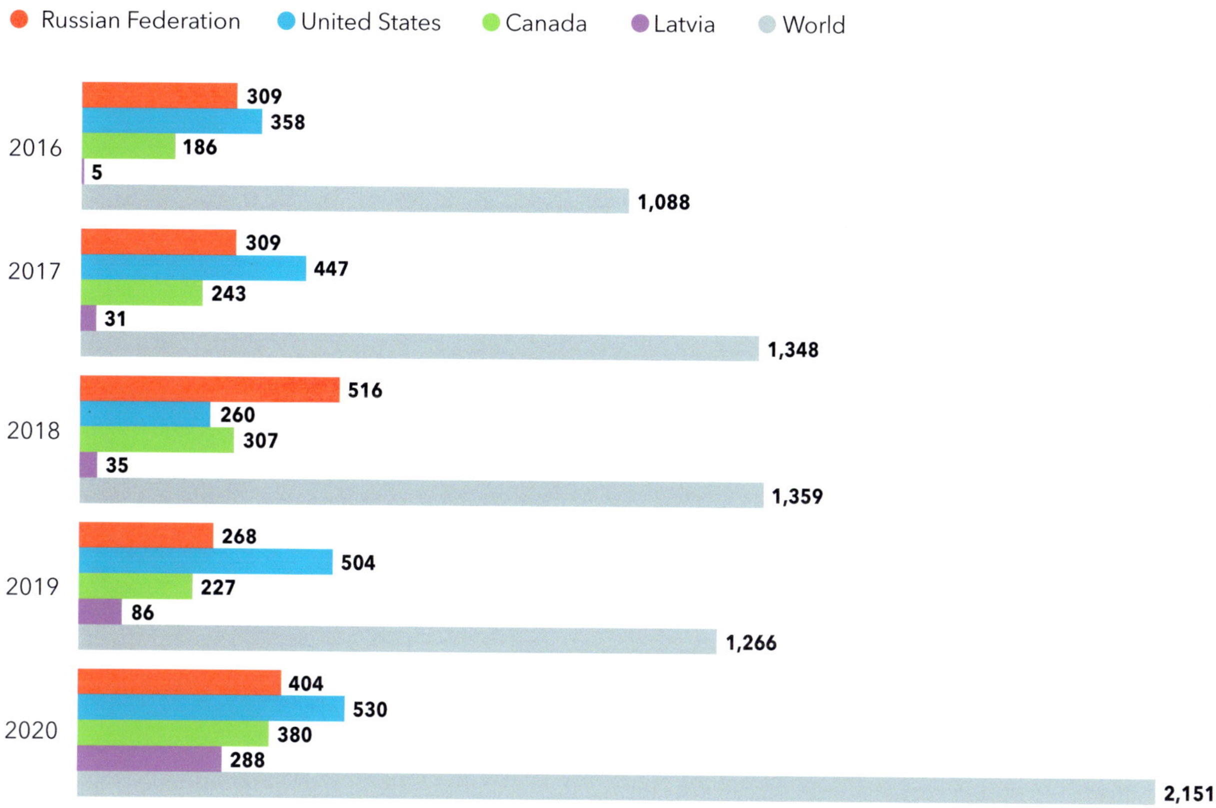

Source: COMTRADE

Nigeria's import dependency is not expected to decline anytime soon, due to fundamental problems such as low adoption of agricultural technologies (which would improve yields), slow progress in agricultural research and development, and shocks such as climate-driven disasters and armed conflicts. For example, the country's production of staple crops including cereals is unable to meet domestic food demand; 99% of the wheat consumed in Nigeria is imported, and of the 7 million metric tons of annual rice consumption 2 million tons are sourced from abroad by smuggling, partly across land borders – despite an official ban on importation of rice in Nigeria.

Thus, the country's food security was vulnerable to global food price volatility even prior to the onset of the COVID-19 pandemic and the recent Russia-Ukraine crisis. Underlying this vulnerability is Nigeria's continued reliance on prohibiting and restricting imports to encourage domestic food production, which typically leads to higher prices of both imported and local food products (World Bank 2022).

Rising wheat prices and their implication for food security in Nigeria

Rising prices in the wake of the pandemic have been a major problem. Data from the World Bank/NBS Nigeria–COVID-19 National Longitudinal Phone Survey 2020 show that food prices experienced a rapid climb immediately after the onset of the pandemic. In March 2021, price inflation of basic food commodities hit 23%, the highest level over the previous two years (NBS 2021), and recent data from Nigeria's National Bureau of Statistics (NBS) show food price inflation continued into March 2022.

Recent findings using comprehensive and long-ranging monthly food price data has shown significant increases in prices for all selected food items during the pandemic, for example, prices of imported rice and wheat increased by 41% and 21%, respectively (Amare et al. 2022).

Figure 2 shows the evolution of wheat prices nationally and averaged across the six geopolitical zones in Nigeria. Wheat price levels and dispersion before the pandemic were relatively stable. Nationally, the price of wheat increased by 21% and the regions experienced a significant increase in price dispersion across markets after the pandemic began, and prices continue to rise.

Figure 2: Price of wheat in Nigeria, 2016–2021, naira per kilogram

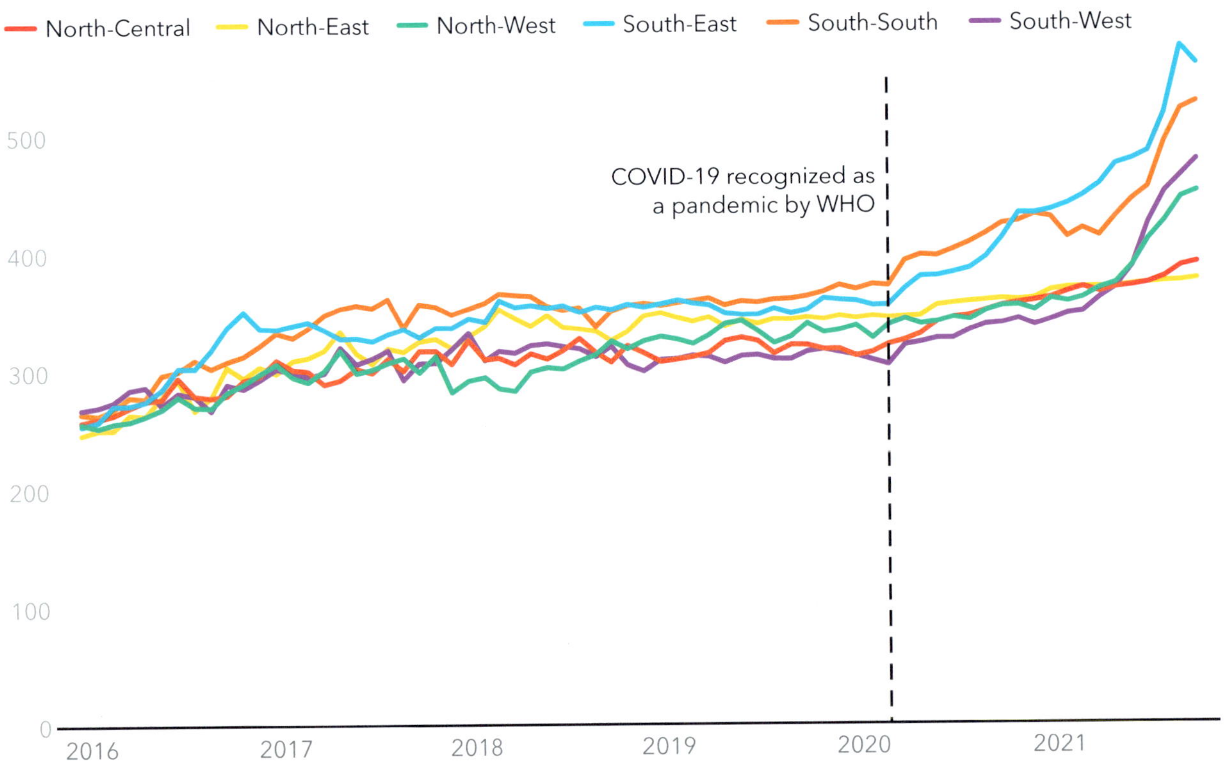

Source: NSI

Wheat flour is, of course, the major ingredient in bread and other confectionaries such as pasta, noodles, semolina, and other foods that are pantry staples in Nigeria. While consumption of these products is higher in urban areas due to better access to markets than rural areas, bread remains an important staple nationwide. The prices of wheat flour – relatively stable up to 2019 – rose by about 3% as the COVID-19 pandemic hit and then by 28% later in 2020. Since then, domestic wheat flour prices have continued to increase both on a year-on-year and month-on-month basis (Figure 3). A similar trend can be seen for bread, both sliced and unsliced; prices were flat before the COVID-19 pandemic but have been rising ever since.

Figure 3: Prices of wheat-related products in Nigeria, in naira, 2017-2020

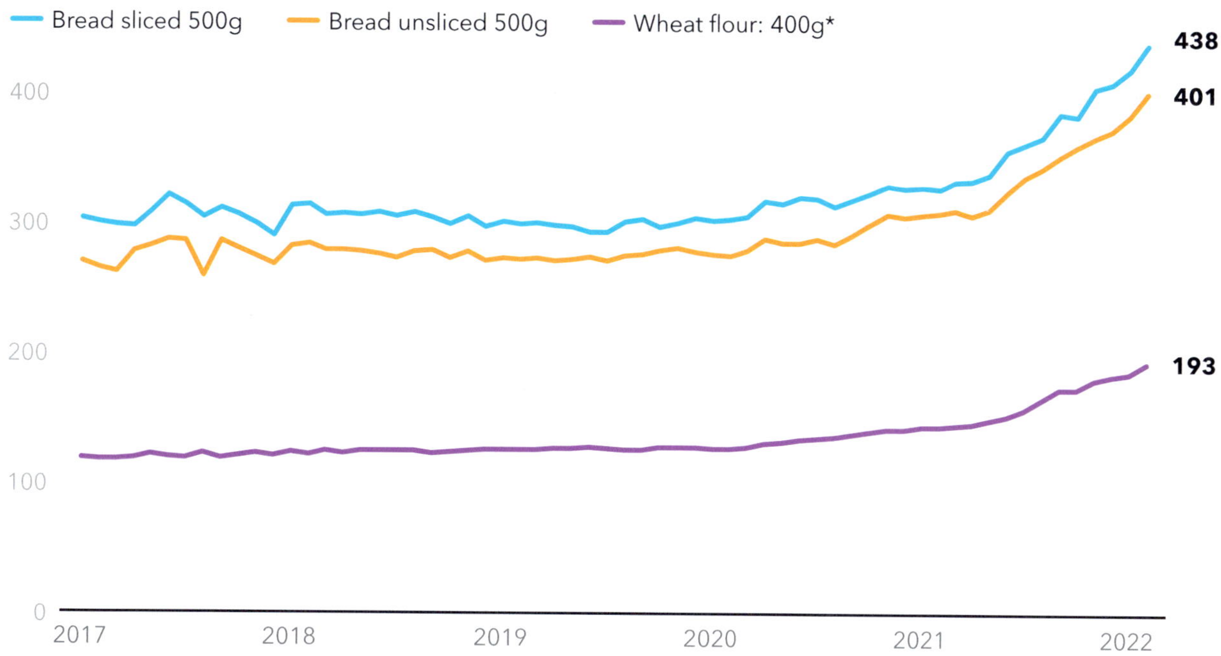

Note: * based on the price of a 2kg prepacked wheat flour bag (golden penny). 400 grams is a proxy for the amount of flour needed to ba 500 gram loaf of bread.

Source: NIS

Now, war-related global market disruptions and rising prices will lead to product scarcity and bring about still-higher domestic prices in Nigeria. Consumers could adjust their consumption patterns and switch to cheaper substitutes such as cassava products; producers may have to lower product quality by mixing wheat flour with alternatives such as sorghum and corn flours to keep prices stable and remain competitive in the market. This will affect food quality and safety and could have negative nutritional implications.

As the country seeks to reduce its overdependence on wheat imports, it may be timely to revisit the "cassava inclusion" policy initiated in 2002 under President Olusegun Obasanjo, which advocated substituting at least 10% high-quality cassava flour for wheat in bread- and confectionery-making.

While the policy failed in the past due to noncompliance by flour millers and other factors, a version of it could work under current conditions.

Given Nigeria's urgent need to reduce dependence on wheat, now may also be time reconsider at least temporary removal of the 15% levy on wheat grain imports introduced in 2012 to drive up local cassava production to replace w wheat flour with cassava flour in bakery production and promote domestic production of wheat.

Fertilizer – an export opportunity

Even as the food security situation looks increasingly worrisome, the crisis may create opportunities for the Nigerian fertilizer industry. Nigeria's annual domestic fertilizer demand is about 700,000 metric tons of nutrients (FAOSTAT); in 2019, the country produced 1 million metric tons per year with 61 fertilizer production and blending plants. That situation is now likely to change with the inauguration in March 2022 of the Dangote Fertilizer Plant – whose production capacity is 3 million metric tons of granulated urea per year. Nigeria is now the leading producer of fertilizer on the African continent. At the plant's inauguration ceremony, President Muhammadu Buhari stated that Nigeria will begin exporting fertilizer.

With Russia among the top exporters of fertilizer in the world, and the conflict and sanctions disrupting global markets, importing countries are looking for alternative sources. Nigeria's timely investment in the fertilizer sector positions it as a suitable alternative and a regional fertilizer hub, which could help reduce some of the negative impacts of the crisis on fertilizer importing countries, particularly those in West Africa.

Natural gas – an economic and social welfare opportunity

Europe's quest to find alternative energy markets and reduce its overdependence on Russian oil and gas also presents an opportunity for Nigeria. The African Energy Chamber Outlook Report (2022) estimates that Nigeria's gas production will increase from 1,450 billion cubic feet (bcf) in 2021 to 1,780 bcf in 2022, improving domestic energy security and facilitating increased exports to European and other markets.

Nigeria is also a current supplier of liquified natural gas (LNG) to several European countries. Now, with partners Niger and Algeria, it plans to construct the Trans-Saharan Gas Pipeline, which will enable transport of LNG from Nigeria to European markets, allowing Nigeria to increase exports and Europe to diversify its natural gas sources.

However, as Nigerian export revenue is highly dependent on crude oil (almost 90% of all exports) and the country's international debt service payments are increasing over time (Central Bank of Nigeria), future investments in the natural gas industry will depend on the prudent fiscal and debt management policies of the country – by no means guaranteed given past experience, but still within reach. Nigeria's debt-to-GDP ratio, around 32% in March 2022, is within a sustainable debt servicing range, and relatively low compared to other countries in West Africa. Thus, whether these potential opportunities amid the Russia-Ukraine crisis ultimately help to solidify Nigeria's position as a major gas exporter to global markets remains to be seen.

Safety nets

Given the rapidly unfolding challenges of the Ukraine-Russia crisis, many Nigerians may find themselves facing increased food insecurity in the coming months.

The government has not currently increased safety net support in response to the Ukraine crisis. One potential source of assistance going forward is the Food Security Cluster (FSC) program for Nigeria, formed by a broad partnership of international organizations. During the pandemic, the FSC provided assistance to qualifying Nigerians in various forms including cash, mobile money transfers, and paper and electronic vouchers. In December 2020, the FSC assisted over 4 million people.

Expanding natural gas production could also provide an opportunity to expand social safety nets. The government could establish a "Natural Gas Fund" from the export revenue to back programs that cushion vulnerable citizens amid critical economic shocks. Similar approaches have been applied in other countries such as Ghana, where the Ghana Petroleum Funds channel some excess petroleum revenue, "serving as an endowment to support future development."

Policy implications

Nigeria must also address its dependence on food imports and other structural problems, and build long-term resilience to food crises and other shocks. It can do so by focusing on the following three key policy areas:

First, to increase domestic agricultural production, the government can accelerate the implementation of the new National Agricultural Technology and Innovation Policy (NATIP 2022–2027), focusing on promoting technologies to improve agricultural productivity, particularly for wheat and other major food grains. The NATIP relies on strategies such as rapid mechanization, the establishment of an Agricultural Development Fund, a revived extension strategy, livestock and fisheries development, market development, and strengthening value chains.

To effectively address both the short-term crisis and longer-term problems, we recommend that the NATIP implementation begin with a priority-setting exercise that estimates the returns to various policy options in terms of poverty reduction, agricultural productivity improvements, and other pressing goals.

Second, the government should reform distortionary agricultural trade policies. The recent reopening of land borders is a step in the right direction. Such reforms would reduce the vulnerability of Nigerian households to food price increases by enhancing the role of markets in moderating supply disruptions. Addressing trade restrictions is also key to Nigeria's participation and leadership in the implementation of the African Continental Free Trade Area (AfCFTA) agreement.

For example, given recent inflation, it may be time to reconsider import restrictions on food items such as rice and chicken meat. A number of important food items – including maize, sugar, and milk – remain on the list of prohibited or restricted items or are ineligible for foreign exchange for imports.

Third, this post has highlighted some areas where there may be positive outcomes for Nigeria from the global crisis. Preliminary results from IFPRI's modeling of the impacts of current global market shocks suggest that Nigeria's economic growth in 2022 will experience positive outcomes from natural gas price increases, with countervailing negative impacts from food and fertilizer price increases. Strengthening the resilience of the economy calls for capitalizing on Nigeria's comparative advantages, exploring export markets (such as fertilizer), and investing in infrastructure development to optimize the production of resources such as natural gas, to make the best of emerging economic opportunities.

33 Rising commodities prices driven by the Russia-Ukraine crisis threaten to undermine Kenya's economy, increase poverty

Clemens Breisinger , Xinshen Diao, Paul Dorosh, Juneweenex Mbuthia, Lensa Omune, Edwin Ombui Oseko, Angga Pradesha, and James Thurlow

Originally published June 10, 2022.

Much of the early attention to the Russia-Ukraine conflict's food security impacts has been concentrated on countries highly dependent on wheat imports from the Black Sea region. Given the important role that wheat products play in the diets of people in Egypt, Sudan, Yemen, Lebanon, and other countries, the interruption in Black Sea wheat trade and high prices have raised serious concerns about rising levels of food insecurity, poverty, and instability around the world.

But many countries are affected by price increases across a range of commodities (some predating the war), including in fertilizers, edible oils, and maize, as well as oil, natural gas, and other energy products. How are these sharp international price increases impacting countries and people, and how can countries respond? Our recent modeling study focusing on Kenya suggests higher prices, particularly for fertilizer, will reduce GDP growth and increase poverty rates in the country, putting an estimated 1.4 million additional people below the poverty line.

Prices in Kenya are on the rise

Inflation in Kenya hit 7.1% as of May 2022, mainly driven by a combination of high world commodity prices and a depreciating exchange rate (Central Bank of Kenya). The majority of maize, rice, and wheat in Kenya's domestic market – including maize and wheat flours – are produced locally, while about 45% of edible oils are imported. Fuel and other petroleum products are almost all imported. Depending on the year, about 30–50% of wheat grains processed in Kenya for final consumption or used for food processing industries come from Russia and Ukraine, and imported edible oils come mostly from Indonesia and Malaysia (UN COMTRADE database, 2022). While almost all chemical fertilizers are imported in Kenya, a relatively small portion come from Russia – about 9.3% of the total value of imported fertilizers in 2021 (UN COMTRADE). Kenya's petroleum product imports come mainly from the Middle East.

Rising prices negatively affect Kenya's economy and poverty

Almost all the commodities affected by recent global crises are consumed directly by households and used for other sectors' production as intermediate inputs. Using IFPRI's economywide Rural Investment and Policy Analysis (RIAPA) model for Kenya, we assess the likely impacts of the ongoing global crises on the Kenyan economy and people by taking into consideration cross-sector linkages and impacts on various supply chains.

The study indicates that Kenya's gross domestic product (GDP) could be 0.8% lower in 2022 compared to a situation without the global commodity price crises. More than 50% of GDP losses are from agriculture, driven primarily by higher fertilizer prices and the associated negative productivity effect from farmers' reduced use (Figure 2). (The study assumes that a 100% increase in real fertilizer prices leads to a 15% decline in fertilizer use and that farmers who do not use chemical fertilizers are about 20% less productive than farmers who use chemical fertilizers.)

Household consumption is also negatively affected by rising world prices – particularly higher fuel prices that raise market prices for many consumer goods and services. The decline in consumption is larger for households toward the lower end of the income distribution, leading to increased inequality and greater poverty.

According to the most recent Kenya Integrated Household Survey Report (KNBS 2018), the overall poverty headcount rate was 36.1%, with a rural poverty rate

Figure 1: Share of imports and domestic production in total commodity supply in Kenya, 2019

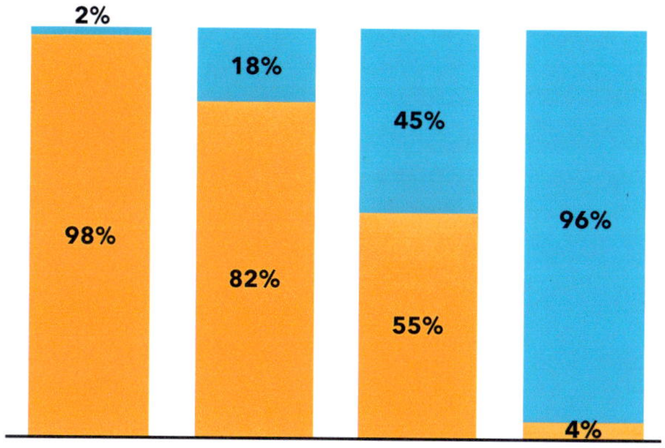

Source: Author's calculations using social accounting matrix (SAM) data from IFPRI's Kenya RIAPA model.
Notes: Maize and wheat include maize flour and wheat flour, and edible oils include edible oilseeds. Final use includes private and public consumption and gross capital formation

Figure 2: Change in total and agrifood system GDP due to food, fuel, and fertilizer shocks

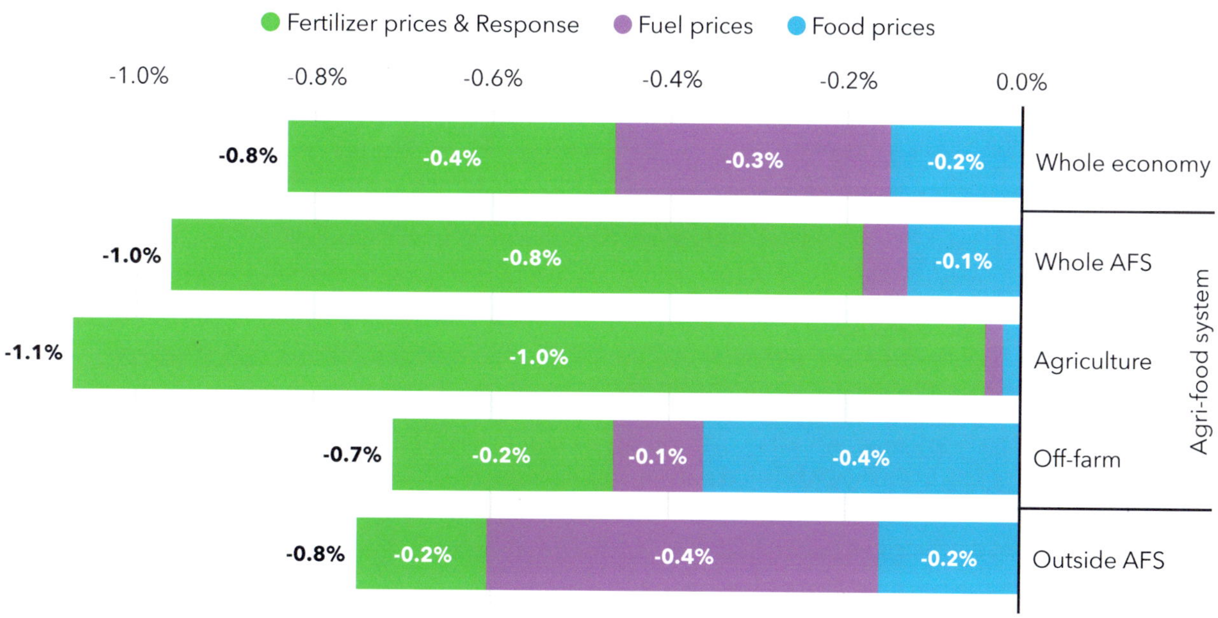

Source: Simulation results from IFPRI's Kenya RIAPA model
Note: About 16% of the effect on agriculture GDP under "fertilizer prices and response" is directly from rising fertilizer, while the remaining 84% is from the productivity shock caused by lowering the use of fertilizer

of 40.1% and core urban rate of 29.4%, in 2016. (The poverty line is defined as individual expenditures of less than $31.90 per month in rural and peri-urban areas, and less than $58.80 in core urban areas).

The increase in world prices raises the national poverty headcount rate in Kenya by 2.5 percentage points, while the rural poverty rate rises more, by 3.1 percentage points, compared with just 1.4 percentage points for the urban poverty rate (Panel A in Figure 3) – this is equivalent to an additional 1.4 million people falling below the poverty line (Panel B in Figure 3). Most of the increase in poverty is caused by the fertilizer shock, and the absolute increase in the poor population is larger in rural areas, where more poor people are concentrated.

Figure 3: Changes in poverty due to food, fuel, and fertilizer shocks

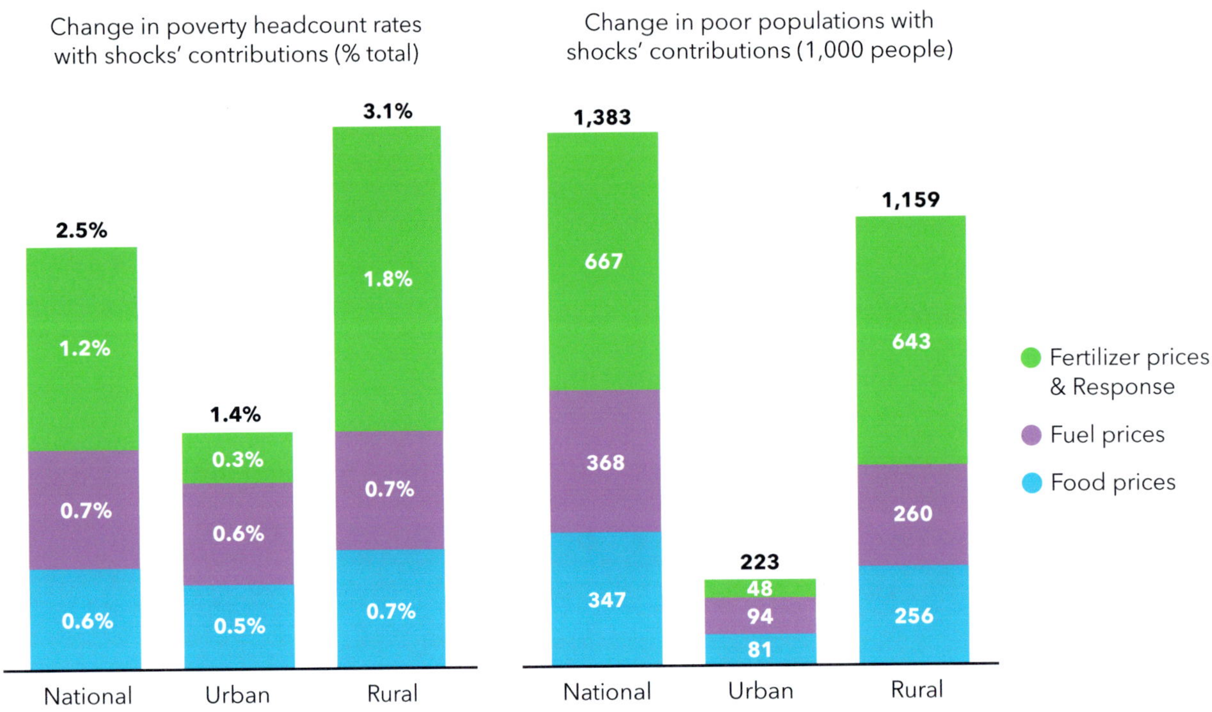

Source: Simulation results from the survey-based microsimulation module within IFPRI's Kenya RIAPA model

The Kenyan government responds with fertilizer and fuel subsidies

To cushion consumers and producers from the impacts rising global prices, Kenya's government has implemented several fiscal measures, including fertilizer and fuel subsidies. As of April 2022, the government had budgeted a total of KSh.5.7 billion (US$49.6 million) to subsidize 114,000 metric tons of fertilizer for crops. As the planting season was already ongoing, anecdotal evidence suggests that farmers largely made their planting and fertilizer adoption decisions before the subsidy mechanism was fully implemented.

The government introduced price controls in the energy sector in 2011, and in 2020 established a stabilization fund that became operational in April 2021. By April 2022, the government had already paid KSh.49.2 billion ($425.8 million) to oil marketers to keep the fuel prices in the country at a reasonable margin. With persistently high oil prices, the current fund is expected to fall short in maintaining fuel prices at pre-Ukraine war levels.

Policy options

In the short run, fertilizer and fuel subsidies can partially blunt the effects of global shocks on Kenya's producers and consumers. Given that global fertilizer prices will likely remain high creating fiscal constraints, more targeted fertilizer subsidy schemes may be necessary before the next planting season. Maintaining the timely availability of fertilizers is also important.

However, going forward Kenya will need to expand its domestic fertilizer production to reduce its dependence on imports. The government has pursued this strategy by allowing the private sector to import fertilizers and blend in the country for specific crops. The government should also consider encouraging private sector investments to establish fully-fledged fertilizer production in Kenya and/or the broader East African region.

Given the impacts of the current crisis on the poor, social safety nets should be temporarily expanded, including existing cash transfer programs such as Orphans and Vulnerable Children (OVC) cash transfers, Persons with Severe Disability (PWSD) cash transfers, the Hunger Safety Net Program, and nutrition-sensitive cash transfers. The efficiency of these programs can also be improved, including meeting challenging problems of late disbursement of funds and delays in targeting and enrollment of beneficiaries, among other issues.

The current global crisis underlines the longer-term urgency of moving toward a green economy transformation. Currently, 75% of Kenya's electricity is generated from hydro, wind, solar, and geothermal energy sources, and there are significant opportunities to expand production and foster green growth in the industry and services sectors.

Expanding agricultural production and productivity, meanwhile, will require an increase of public and private investments in agriculture, improvements in the business climate, and more effective implementation of agricultural and related investment projects under the government's Vision 2030 and Big 4 Agenda.

Our study indicates that rising prices of key commodities sparked by the Russia-Ukraine war are stressing Kenya's economy, with the poor bearing the brunt. A combination of short- and longer-term actions by the government, business, and international partners can help to cushion those impacts, and strengthen the country's economy against future shocks.

Update

Since this analysis was published in June 2022, Kenya has elected a new government under the leadership of President William Ruto. At the core of the new government's development agenda is the Bottom-Up Economic Plan, with the main objectives of creating jobs, eradicating hunger, and lowering the cost of living for its citizens. Several of the Plan's core elements such as the focus on agriculture and the poor, revamping of fertilizer subsidies and distribution, and the longer-term goal of moving toward a green economy transformation were recommended in our original analysis. Responding to a request from the Kenyan National Treasury and Economic Planning and in collaboration with the Kenya Institute for Public Policy Research and Analysis (KIPPRA), IFPRI staff have expanded the Kenya RIAPA model used for the Ukraine-Russia impact analysis to also analyze the expected impacts of implementing the Plan from 2023 to 2027.

Analysis of the impacts of the Russia-Ukraine war and the Bottom-Up Economic Plan were conducted by rapid response teams composed of Kenyan experts and institutions and IFPRI researchers under the CGIAR Initiatives on National Policies and Strategies and Foresight and Metrics. This collaborative approach to research has greatly enhanced the relevance and uptake of the findings and appreciated by Kenyan policymakers.

This post was drawn from a recent study conducted by IFPRI with financial support from the Bill & Melinda Gates Foundation (BGMF), the UK Foreign, Commonwealth & Development Office (FCDO), and the U.S. Agency for International Development (USAID). The study uses models developed with ongoing support from BMGF, USAID and the CGIAR Foresight and Metrics initiative. The study also benefited from collaboration with the CGIAR National Policies and Strategies (NPS) Initiative and IFPRI's Kenya country program. Thanks to Joseph Karugia, ILRI, ONE CGIAR Country Convener for Kenya, for facilitating the NPS technical and the representatives from the Ministry of Agriculture, Livestock, Fisheries and Cooperatives, Kenya Agricultural and Livestock Research Organization, Tegemeo Institute, Kenya Institute for Public Policy Research, Egerton University, and private sector representatives and other stakeholders for attending this NPS technical meeting and for their valuable inputs.

34 How the war in Ukraine threatens Bangladesh's food security

Abdullah Mamun, Joseph Glauber, and David Laborde

Originally published April 20, 2022.

As the Russia-Ukraine crisis continues to disrupt the global trade of key foods such as wheat and vegetable oils, along with fertilizers, impacts are falling heavily on countries such as Bangladesh. Dependent on imports of those items to feed its large population, many of whom are poor and vulnerable to shocks, the country faces the prospect of rising food insecurity.

The conflict threatens Bangladesh's recent progress on this front. With a population of 165 million in 2021 – with 38% employed by the agriculture and fisheries sector – and a growing economy, the country's undernourished population declined from a high of 16% in 2000 to as low as 9.7% in 2019. While the COVID-19 pandemic increased food insecurity, by some measures the country proved relatively resilient: According to an IFPRI study, the proportion of rural households facing moderate or severe food insecurity rose from 15% in early 2020 to 45% in January 2021, then returned to pre-pandemic levels by the end of 2021.

Now that 2021 recovery is in danger: Bangladesh saw a record rise in prices of staples in March 2022, along with volatility in the fertilizer market. In this post, we discuss Bangladesh's trade exposure to several commodities facing export restrictions, the fiscal impact of rising imports, and potential measures for easing food security pressures.

Risks for the Bangladesh food system

The war in Ukraine has affected Bangladesh in three significant ways.

First, the wheat market. Although Bangladesh historically has a high reliance on rice in its caloric intake (about 68% of total calories), wheat has been an increasingly important food source – reaching about 7% of total calories. Wheat consumption has tripled since 2000, and Bangladesh imports more than 80% of its wheat needs. Since 2005, Russia and Ukraine have accounted for an increasing share of those imports, reaching more than half of the total (Figure 1). With those sources now disrupted, Bangladesh will have to find alternatives. In previous years, Bangladesh has purchased wheat from India and Australia, among other suppliers. India is expecting a good harvest and has indicated it may be able to supply as much as 10 million metric tons to world markets this year.

Second, the vegetable oil market. Ukraine and Russia account for about 75% of world sunflower oil exports. While Bangladesh does not import sunflower oil, the war has increased prices across the entire vegetable oil complex. Bangladesh imports almost all of its vegetable oil either in the form of raw or processed products (primarily palm oil and soybean oil) or imported oilseeds (rapeseed and soybeans), processed domestically.

Trade restrictions imposed by major vegetable oil exporters have also increased Bangladesh's vulnerability. Indonesia supplies 80% of Bangladesh's palm oil imports. In March 2022, Indonesia implemented a progressive

Figure 1: Wheat and vegetable oil imports to Bangladesh, million metric tons per year

A. Wheat imports

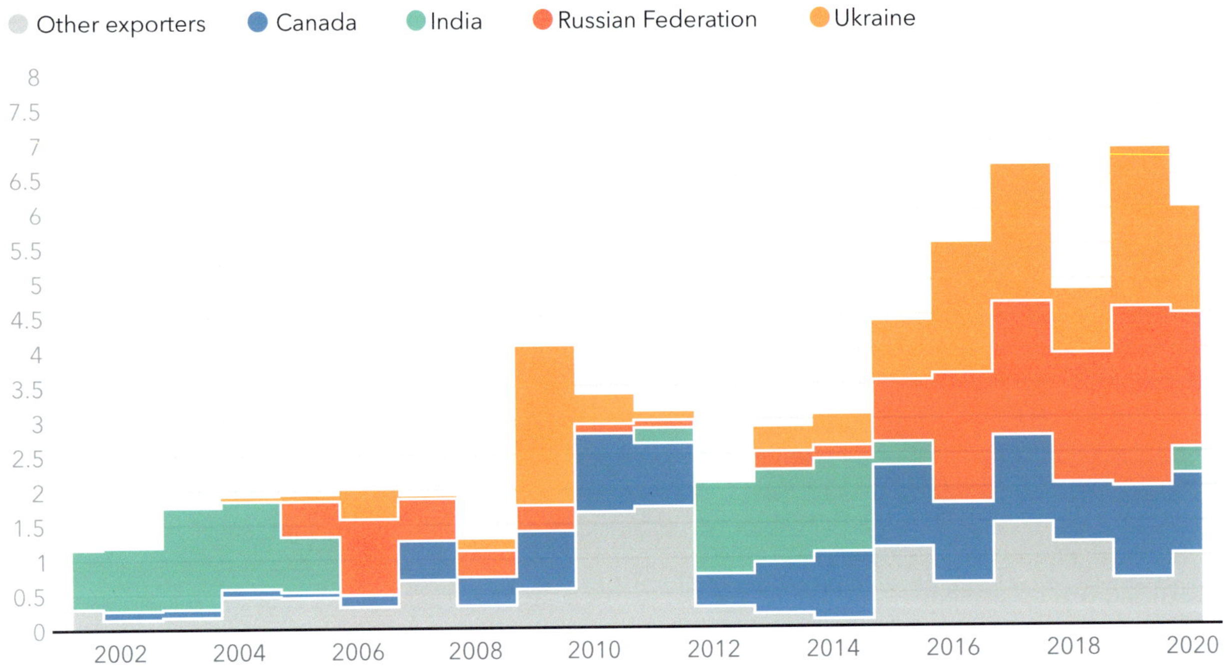

B. Vegetable oil imports

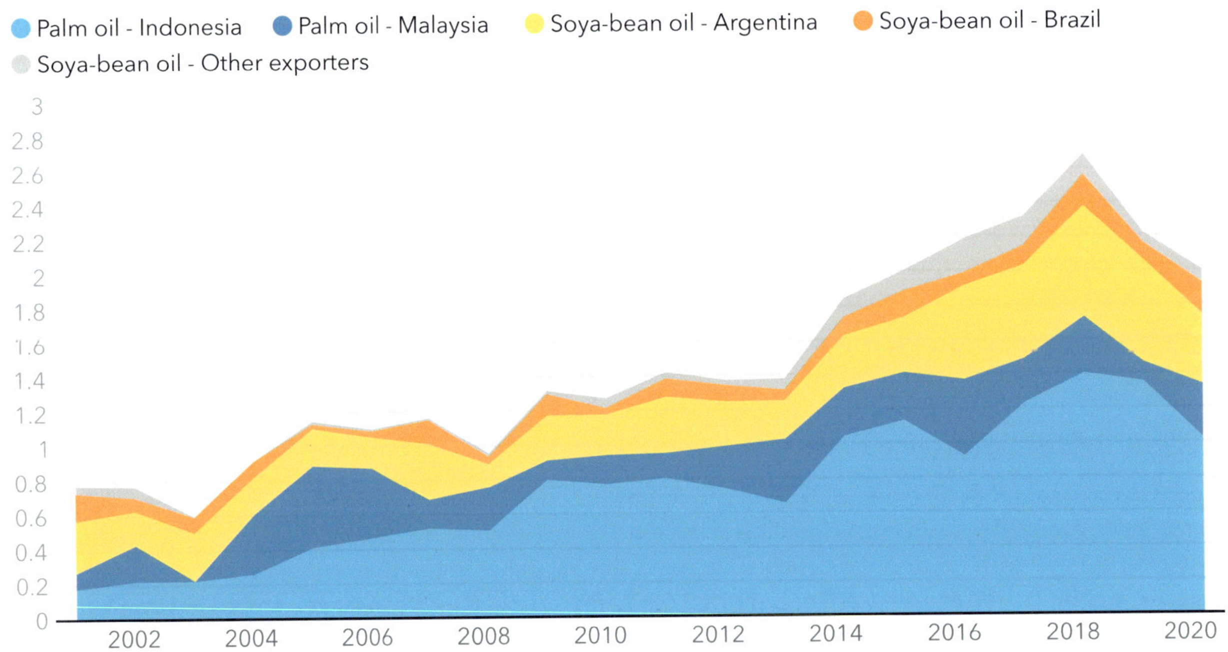

Source: COMTRADE

export levy for palm oil and its derivates that may limit total supply in the global market. Argentina meanwhile accounts for about two-thirds of Bangladesh's soybean imports. On March 13, Argentina restricted exports of soybean oil and meal, then altered course on March 31, permitting exports but raising export taxes on those products from 31% to 33%. All of these factors have kept vegetable oil prices at record levels and increased import costs for countries like Bangladesh. Considering the effects on both grains and oilseed markets, our export restriction tracking data indicates that 38% of Bangladesh's total imported calories is likely to be impacted by such measures, equivalent to a quarter of the dollar value of total imports.

Third, fertilizer costs. Bangladesh's agriculture, particularly rice production, relies heavily on the use of fertilizers. Average application rates are over 286 kg/ha and Bangladesh imports over 1.2 million metric tons of fertilizer annually, including 31% of nitrogen needs, 57% of phosphate needs, and 95% of potash needs. Russia and Belarus are major global fertilizer exporters, and the war, including export sanctions aimed at Russia, has disrupted these markets as well. Bangladesh imports almost 75% of its potash from Russia (34%) and Belarus (41%). Bangladesh will likely have to source these ingredients from other countries and pay higher prices. Resulting lower input use – particularly of nitrogen-based fertilizers – could reduce rice production, increasing demand for rice imports.

In addition, Bangladesh imports 44% of its corn and soybean supplies (computation based on FAOSTAT food balance sheet), and these commodities too are seeing rising global prices. This impacts the feed sector in Bangladesh, increasing the production costs of poultry, cattle, and fish products – sectors already weakened by the pandemic. Reducing the availability of animal products in the coming months could compound the food security impacts from disruptions in the key staple products of wheat, vegetable oils, and rice.

To maintain supplies, Bangladesh will have to buy commodities at higher prices, which will put pressure on monetary reserves and impact the currency market; there have been signs of currency depreciation already. This in turn will lead to an additional price increase for consumers, as high world market prices are amplified by a weaker exchange rate, and more pressure on public finance, trapped between servicing the country's foreign debt (already US$460 per capita in 2021) and the increased cost of domestic programs.

Macroeconomic consequences and direct impact on government policies

As in many countries, domestic food prices are already rising. Though the official estimates of price inflation – 5.5% for food in December 2021 – depict a relatively mild situation, a report by the Centre for Policy Dialogue (CPD) indicates higher prices for essentials (with rice thus far remaining relatively stable).

In the first seven months of the current fiscal year (the fiscal year begins in July), the trade deficit has increased by over 82% to $18.7 billion, as the import bill surged by 46% over the same period in FY21, and despite high export growth. Costs include $5.79 billion for rice, wheat, fertilizer, and vegetable imports. According to the U.S. Department of Agriculture, Bangladesh is expected to import 7.5 million tons of wheat and 2.3 million tons of corn in 2022/23.

How can Bangladesh cushion impacts of rising prices and supply problems on the poor and on farmers? The government runs extensive public food distribution programs (PFDS) including Open Market Sale (OMS), the Food Friendly Program (FFP), and the Vulnerable Group Feeding (VGF) program. The FFP program alone provides food, mainly grains, to over 5 million poor families with an annual cost of $380 million. During the first year of the pandemic, public food distribution rose to 2.8 million tons from the pre-pandemic level of 2.6 million tons (2018/19). In the current fiscal year the country has already distributed 1.5 million tons against a target of 3.3 million tons of food grains.

Figure 2: Bangladesh's food and fertilizer import bill, US dollars billions

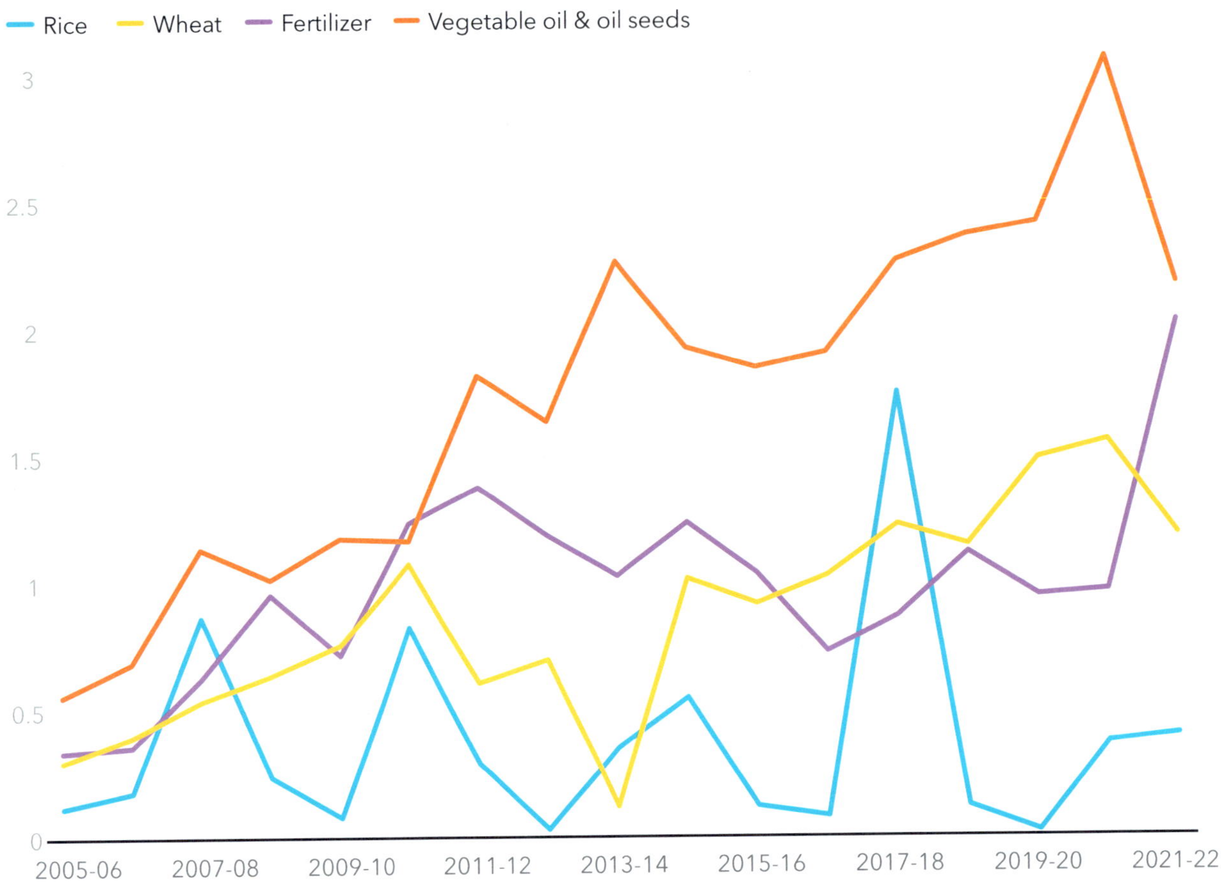

Note: 2021-22 represents Jul to Dec of 2021 only.

Source: Bangladesh Bank

Figure 3: Public food distribution gains in Bangladesh, million metric tons

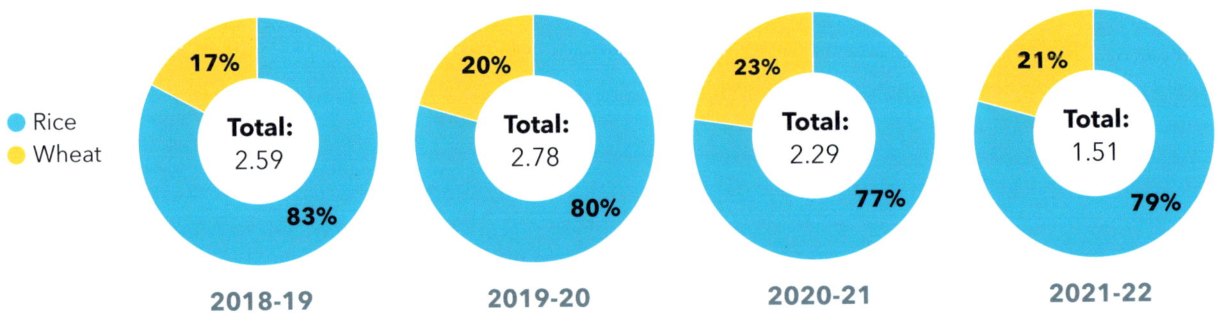

Note: Year represents fiscal year (July-June) 2021-22 data as of Dec 2021

Source: Ministry of Food, Bangladesh

The greater volume of the distribution comes as income losses and high unemployment due to COVID-19 in the recent months have pushed many families into poverty, leading the government to expand food distribution programs, particularly through OMS. The decision in March 2022 to provide another 10 million families with soybean oil, sugar, chickpeas, and lentils at subsidized prices is a clear indication of the country's growing food insecurity problem.

Meanwhile, government subsidies for fertilizers are expected to hit a record high $3.2 billion in 2021/22, tripling the initial allocation of $1 billion. Farmers receive subsidies that account for 70% to 85% of market prices of various types of fertilizers. With a total demand of 5.8 million tons of fertilizers, and market disruptions proliferating, farmers now face huge uncertainties in getting fertilizers on time. Fertilizer shortages would have significant impacts in the planting of Aus season rice this spring, Aman in July, and Boro season rice later this fall.

Policy recommendations

Food security for the poor is at the core of the government's policy agenda. Though the country's production forecast for rice (36.3 million tons in 2022/23, according to USDA) gives some level of comfort, it still needs to procure 0.7 million tons of rice from abroad. International wheat and maize supplies are uncertain, as are those of fertilizers. How can Bangladesh cope with the current price and supply volatility in these items and maintain food security?

In order to keep the inflationary pressure under control, the government should monitor price trends in both domestic and international markets and take timely measures to maintain sufficient food supplies.

Research by Chowdhury and colleagues (2021) shows that there were leakages in the FFP distributions during the early phases of the pandemic, due to mis-targeting recipients and in appropriations of the quantity disbursed, among other issues. Since the government has now decided to increase the number of people receiving food, it is crucial that the selection of beneficiaries be done properly, preferably by automating the process. Routine monitoring and effective evaluation of such programs can also help alert the government about supply disruptions and inefficiencies. Similar approaches should be used for fertilizer programs to improve their efficiencies.

To help stabilize markets, some short-term initiatives would be helpful. The government should continue to provide assistance to consumers through OMS and other mechanisms. In this regard, we find it encouraging that the government has already reduced import tariffs and the value added tax (VAT) to keep prices low. To address structural weaknesses, longer-term reforms will be required – in particular, a reform of fertilizer subsidies, as such policies are not sustainable from an environmental and fiscal point of view.

Conclusion

As the ongoing Ukraine-Russia conflict threatens global food security, Bangladesh faces a predicament: paying rising prices for import commodities to maintain supplies will squeeze currency reserves further and leave little room for the government to expand social safety net programs.

Meanwhile, global market problems outside of food also threaten the country's economy. The textile and garment sector could be affected, putting income and jobs at risk for many poor households. Indeed, input costs have been rising, both for synthetic and natural fibers (+72% increase in cotton prices over the last two years), while global demand has fallen during the pandemic, and the Russian market for Bangladesh's exports now faces uncertainties in delivery of goods and payment receipt due to the war and international sanctions.

These concerns are already undermining Bangladesh's long-term recovery from the COVID-19 pandemic and will no doubt delay progress in achieving the SDGs by 2030. A coordinated approach by the government, with the support of international partners like the IMF to ensure proper financing, and combining consumer subsidies, market intelligence, and quick evaluation of food and fertilizer distribution program performance, is needed to effectively manage Bangladesh's food security needs.

Update

Bangladesh has been facing one of the most severe macroeconomic challenges in decades as its currency reserves were quickly depleted in recent months. Driven by the rising prices of key imported commodities, the trade deficit that stood at US$33 billion at the end of fiscal year 2021/22 has continued to rise while exports and remittances declined in July–September 2022. Due to sharp depreciation of the domestic currency and global commodity spikes, the country has been experiencing very high prices for essential food items. Prices of gasoline and diesel have also risen rapidly, adding immense economic pressure for many people. Rising prices have increased the government's costs for food, fuel, and fertilizer subsidies, which have jumped to $15.4 billion, three times the original allocation in the 2022/23 budget. To avoid fiscal and monetary stress, the government is in negotiations with the IMF for a potential $4.5 billion loan package. Government-to-government discussions are being held regarding rice and wheat imports, particularly with India, Viet Nam, and Thailand, to avoid food shortages in 2022/23. Several challenges lie ahead in ensuring food security in the coming year – maintaining fertilizer subsidies; providing diesel and electricity for irrigation; managing soaring import bill for staples, including grains and vegetable oils, at the expense of foreign reserves; and importing fertilizers from alternate sources in time, as the traditional sources, including Russia and Ukraine, are absent.

35 Can agricultural exports from Southern Cone countries make up for global supply disruptions arising from the Russia-Ukraine war?

Joseph Glauber, David Laborde, Valeria Piñeiro, and Agustín Tejeda

Originally published November 14, 2022.

The economies of the Southern Cone (Argentina, Brazil, Chile, Paraguay, and Uruguay), major agricultural exporters still recovering from the negative impacts of the COVID-19 pandemic, have benefited from the rise in international prices accompanying the Russian invasion of Ukraine. With the war disrupting global supplies, the net exporting countries of the region have the opportunity to increase their exports. If these countries, especially Mercosur members, can boost production enough to offset the loss of Ukraine exports, it would go a long way toward stabilizing markets and world food security.

However, like the rest of the world, these countries are operating in the difficult current environment of global economic slowdown, rising energy and fertilizer prices, and rising inflation. In this post, we examine responses of Southern Cone countries to the disruptions in agricultural markets caused by the war, and their prospects going forward.

The pre-war situation

To understand the effect of Russia's invasion on world markets, it is important to consider overall market conditions prior to the invasion.

Global stocks for principal grains and vegetable oils were at the lowest levels in years as a result of post-pandemic demand recovering much faster than anticipated, especially in China, which reached record import levels for the majority of its agricultural products. Additionally, in South America, the La Niña phenomenon led to lower-than-expected levels of production. At the same time, global inflation was rising, along with international freight rates and input prices. Lastly, supply chains – which appeared to be returning to normal – suffered greatly from recurring strict COVID-19 quarantines in China, which continue to cause significant congestion in the region's ports.

Thus, before the conflict erupted, the world needed a large global harvest during the 2022/23 season to rebuild grain and oilseed reserves. The interruption of Ukraine's exports and uncertainty over shipments drove international prices to record levels in nominal terms, disrupting trade and forcing many countries to seek alternative suppliers.

Impact on Southern Cone agricultural exports

With Russia's invasion of Ukraine on February 24, 2022, grain and oilseed prices rose to new highs, building on a two-year rally. While they have returned to pre-war levels in recent months, they remain well above 2020 levels, and price volatility has been at the highest levels since the food price spikes of 2007/08 for wheat and 2012/13 for maize and soybeans.

Wheat, in particular, has suffered from the combination of war-related supply disruptions and 2021 droughts in North America, with importing countries of South Asia (for example, Bangladesh) and the Middle East and North Africa (MENA) facing problems. Wheat is a major part of their populations' diets, so these countries face the risk of political and social crises as prices rise. Thus, finding alternative suppliers was critical – sparking a redirection of trade flows with intraregional impacts (Figure 1).

Figure 1: Change in wheat exports by exporter, Jan.-June 2022 vs. 2021, metric tons

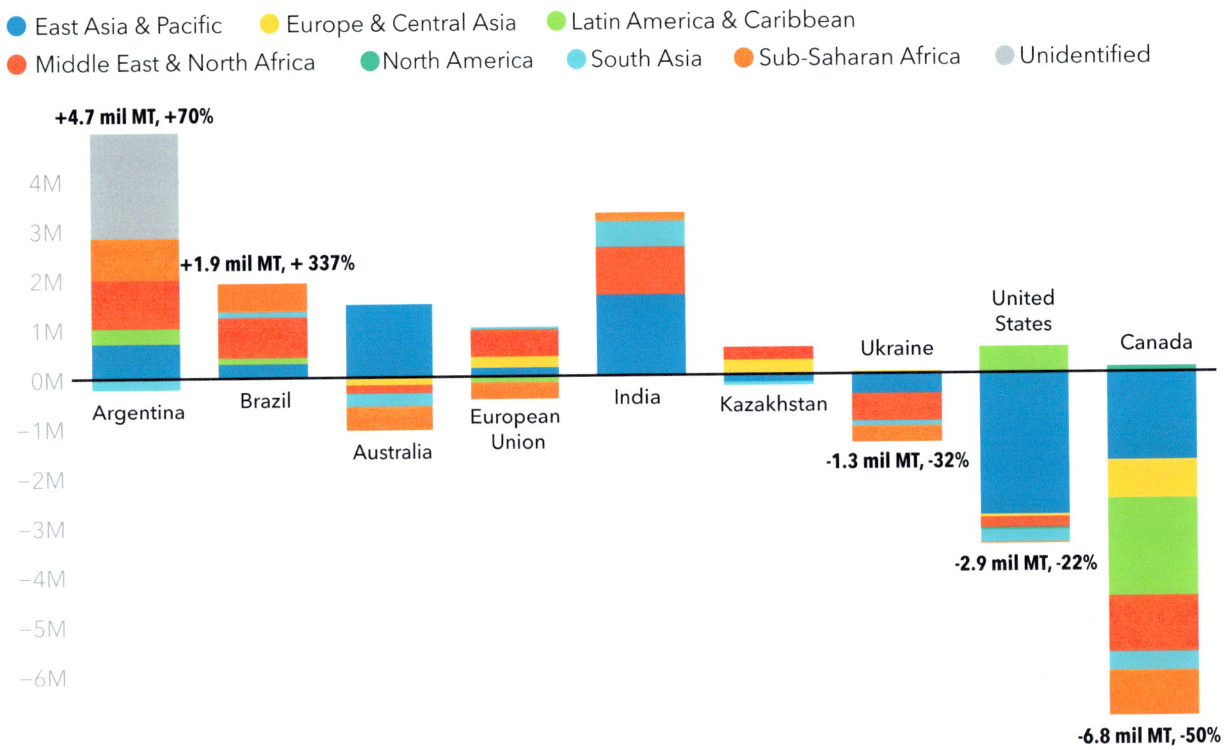

Note: MERCOSUR countries traditionally realize 2/3 of their wheat exports during the first six months of each year.
Source: COMTRADE, TDM

In the first six months of 2022, Argentina and Brazil increased their wheat exports by 6.6 million metric tons compared to the same period the previous year – that figure now amounts to two-thirds of total wheat exports from the two countries. Brazilian wheat is being exported to markets in the MENA region (for example, Sudan, Saudi Arabia) at levels not seen in the last decade. Together, Brazil and Argentina have managed to compensate for the decline in exports from Ukraine, Canada, and the United States to the MENA countries, playing an important role in stabilizing these markets. A similar shift took place in markets in sub-Saharan Africa (SSA), where Southern Cone countries have partially compensated for reduced exports from Australia, the EU, Canada, and Ukraine.

Overall, the share of wheat exported from Mercosur countries to SSA jumped from 4% to 12% in the first six months of 2022 compared to the same period in 2021, and MENA's share doubled from 9% to 17%. At the same time, total Mercosur wheat exports to Latin America and the Caribbean (LAC) also rose, from 4 million metric tons to 4.5 million tons, while the share of LAC in total exports fell from 51% to 31%. Within these shifting trade flows, some traditional neighboring importers have lost some of their share – for example Argentine wheat exports to Chile fell almost one-third from 2021 levels. Nevertheless, the robust expansion in production and exports shows how the Southern Cone can address both regional and global demand.

Maize is a similar story. Exports from Argentina and Brazil rose 6% and 72%, respectively, between January-August 2021 and the same period in 2022 (Figure 2).

Figure 2: Change in corn exports by exporter, Jan.–Aug. 2022 vs. 2021, metric tons

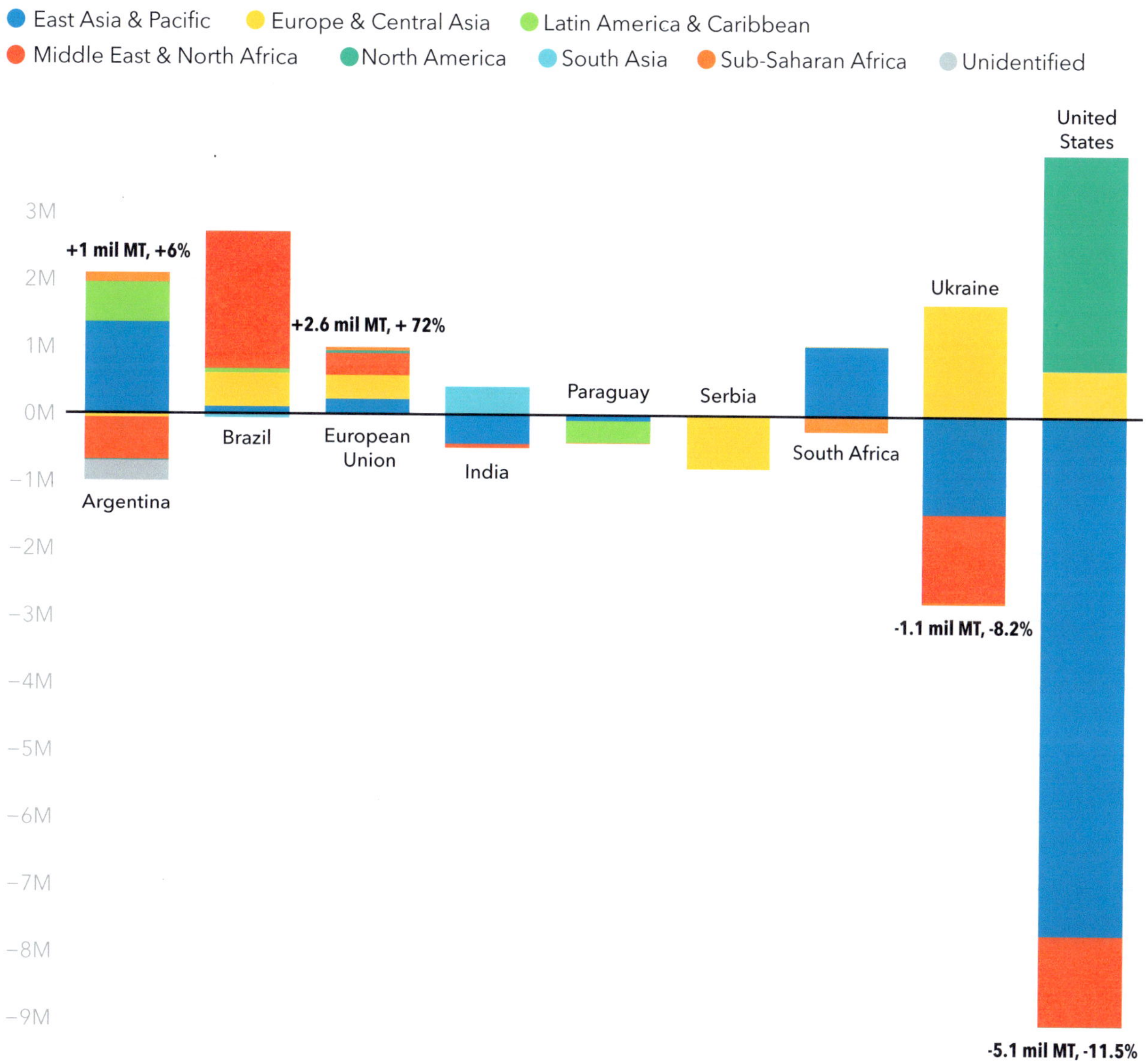

Source: COMTRADE, TDM

Will these trends continue? Unfortunately, drought driven by the La Niña phenomenon is now a serious problem affecting both wheat and maize.

Drought conditions in Argentina, coupled with increased export restrictions, could limit wheat exports in 2022/23. Corn plantings are also affected, and total production is projected to fall by 7.7% (to 48 million tons) in 2022/23 compared to the previous year, according to the Buenos Aires Grain Exchange.

Drought has had some impact on Brazilian maize as well, but so far not affected the country's total corn production, which will likely reach a record figure of 126 million tons according to CONAB data, due to a higher second-crop corn production compensating for losses in the first crop.

Export benefits

Overall, all the Southern Cone countries are net food exporters, so their economies stand to benefit in the aggregate from currently high global prices. Record food exports are estimated for Argentina and Brazil, which would have positive impacts on foreign exchange income.

In the case of Argentina, tax collection via export duties should also increase significantly, bringing the country closer to meeting the goals of its IMF agreement on external debt. The U.S. Federal Reserve's recent series of interest rate hikes has important consequences for emerging economies, driving up the value of the dollar and global commodity prices and increasing the size of debt payments. Southern Cone countries are exposed, given their high debt-to-GDP ratios and dependence on foreign currency provided by primary exports. These conditions are also affecting investment; for example, the Central Banks of Argentina and Brazil have increased interest rates to strengthen their currencies and prevent import prices from skyrocketing.

Production costs for Southern Cone countries: The issue of fertilizers

Despite high global prices, the production and export gap for grains and other products created by the Russia-Ukraine war is substantial. The rest of the world's exporters, including those in the Southern Cone, are not expected to generate enough output growth to fill the whole gap in the coming months.

In addition to drought, increased production costs have put pressure on producers. With global markets disrupted by the war (Russia and neighboring Belarus are large fertilizer suppliers), global fertilizer and energy prices have been at record-high levels.

Among the Southern Cone countries, Brazil has the largest dependence on imported fertilizer, ranging from 71% for phosphates to 99% and 97% for nitrogenous and potash products, respectively.

Argentina, by contrast, despite having higher in-country fertilizer production providing for 35% of domestic nitrogen and 25% of phosphate use, has relatively lower levels of fertilizer application, which translates in some cases to a deficit in nutrient replenishment. Although the country remains 100% dependent on imported potash, its overall relatively lower rate of application gives Argentina a relative economic advantage in dealing with high prices and supply disruptions. But it is at risk of soil nutrient depletion, endangering soil health.

For the first eight months of 2022, potash imports were up for many Southern Cone countries, including Brazil, relative to the same period in 2021 (Figure 3), while the mix of suppliers has changed, reflecting the decline in Russian and Belarusian exports. In 2021, for example, about 51% of the region's potash imports from January through August came from Russia (30%) and Belarus (21%). For the same period in 2022, imports from Russia were roughly the same as in 2021 (30%), but imports from Belarus accounted for just 11%. Countries made up for this deficit by importing more potash from Canada (42% in 2022 vs. 32% in 2021).

Figure 3: Southern Cone potash exports, Jan.-Aug. 2021 and 2022, metric tons

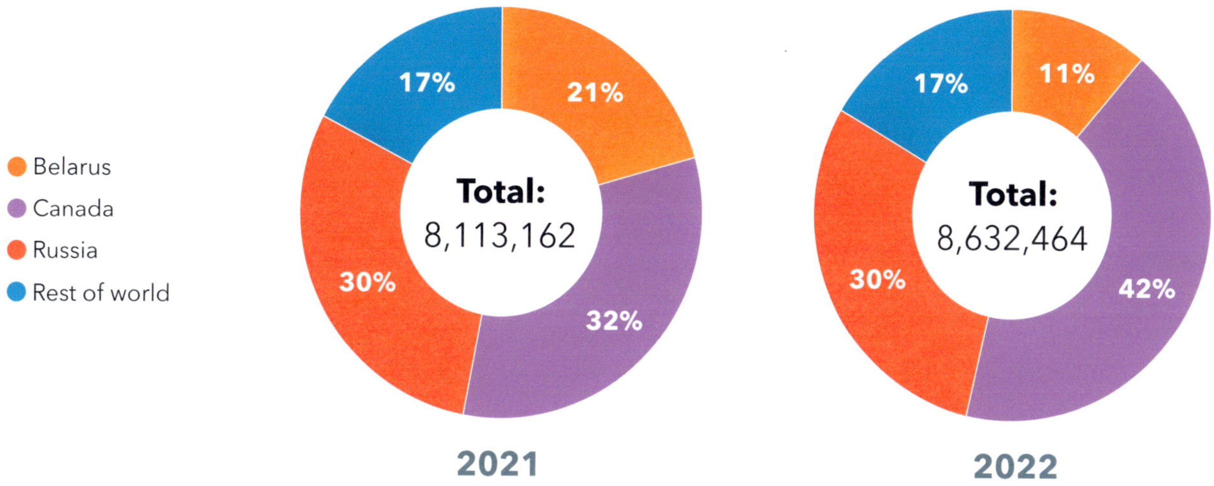

Source: COMTRADE

In addition to affecting total production, increased fertilizer prices encourage producers to choose crops less dependent on input use and with less investment per hectare, such as soybeans.

Moreover, Southern Cone countries, with the exception of Paraguay, which has a surplus from hydroelectric power generation, are paying much higher prices for fuel imports, which represent 15–20% of all merchandise imports of the region. In macroeconomic terms, the increase in the imported value of fertilizers and energy offsets the net gains from rising grain exports in the trade balance and dollar income of the Southern Cone economies. Argentina has responded with greater restrictions on fertilizer imports to prevent foreign exchange outflow.

Increasing the region's fertilizer production in the short term is a major challenge; it might be feasible with investments in the medium term. In the meantime, farmers are likely to continue increasing the use of management technologies such as the variable application of inputs in order to optimize the use of inputs and resources. There is also a greater use of fertilizers of biological origin.

The risk of food inflation, export restrictions

As in previous episodes of high food prices, concern about shortages and inflation creates incentives for governments to implement export restrictions. Such measures disrupt global trade flows, increase market tension, and raise prices. During 2022, food export restrictions were imposed by India, Indonesia, Russia, Ukraine, some Eastern European countries, and Argentina, among others (some have since been rescinded; currently 17 countries maintain some kind of food export ban).

Argentina, in particular, has a long history of applying export duties and restrictions, which resulted in a significant tax on the agriculture sector (representing 19% of farm receipts, according to the OECD PSE indicator). In response to accelerating inflation in August 2021, the Argentine government restricted beef exports, and since December has imposed quotas on wheat and maize exports (though they have not been significantly binding to date).

These measures have little impact on consumer food prices (Calvo 2014), but have had negative effects on producer incentives by reducing domestic producer prices, eroding transparency, and affecting the functioning of markets. These policies not only limit the aggregate response of the agriculture sector, but distort relative prices and influence the selection of crops.

The expectation of greater restrictions may push producers in Argentina to shift toward crops which they see as having a lower probability of intervention, such as soybeans and barley. This is another reason why, despite high prices, Argentina is planting less wheat and maize this summer and fall.

Conclusion

The world faces a continuing food security crisis. The effects the Russia-Ukraine war suggest that it will be difficult for grain, meal, and vegetable oil markets to return to equilibrium in the short term. Current estimates show a further drop in global stocks for the 2022/23 season.

Southern Cone countries have the potential to replace those lost exports from Ukraine and Russia. The region, especially the Mercosur countries, has natural resources, technology, and a dynamic and innovative business model – all of which have increased its resilience in the face of disruptions in fertilizer markets. The region now has the opportunity to cement itself as a reliable global supplier of food, in the face of instability in its main competitors. Taking that opportunity and boosting production and exports would have positive impacts on the economies of the Southern Cone.

Actually doing so will be hard. It will depend on the development of a common agenda of policies favorable to investment and exports. Avoiding export restrictions should be the priority. But that alone is not sufficient; it must be complemented by renewed negotiations to discipline such actions in international forums such as the WTO. As we've seen in 2022, weather shocks continue to pose a major threat to production systems, and investing in resilience strategies is more urgent than ever. The development and adoption of technologies that allow for more efficient use of resources, better environmental performance, and improving soil health is also necessary to protect the long-term capacity of the Southern Cone food system to contribute to regional and global food security.

Contributors

Joseph Glauber is a senior research fellow with IFPRI's Markets, Trade, and Institutions Unit, Washington, DC, USA.

David Laborde is director of the Agrifood Economics Division, Food and Agriculture Organization of the United Nations, Rome, Italy, and a former senior research fellow with IFPRI.

Kibrom Abay is a research fellow and country program leader with IFPRI's Development Strategies and Governance Unit, Cairo, Egypt.

Lina Abdelfattah is a research associate with IFPRI's Development Strategies and Governance Unit, Cairo, Egypt.

Dolapo Adeyanju is a research analyst with IFPRI's Development Strategies and Governance Unit, Abuja, Nigeria.

Mulubrhan Amare is a research fellow with IFPRI's Development Strategies and Governance Unit, Washington, DC, USA.

Kwaw Andam is a senior research fellow and country program leader with IFPRI's Development Strategies and Governance Unit, Abuja, Nigeria.

Bedru Balana is a research fellow with IFPRI's Development Strategies and Governance Unit, Abuja, Nigeria.

Mariia Bogonos is the head of the Kyiv School for Economics Agrocenter, Kyiv, Ukraine.

Antoine Bouët is director of CEPII, Paris, France, and a former senior research fellow with IFPRI.

Clemens Breisinger is a senior research fellow and country program leader with IFPRI's Development Strategies and Governance Unit, Nairobi, Kenya.

Elizabeth Bryan is a senior scientist with IFPRI's Natural Resources and Resilience Unit, Washington, DC, USA.

Joachim De Weerdt is a senior research fellow with IFPRI's Development Strategies and Governance Unit and leader of IFPRI's Malawi Strategy Support Program, Lilongwe, Malawi.

Xinshen Diao is deputy director of IFPRI's Development Strategies and Governance Unit, Washington, DC, USA.

Paul Dorosh is director of IFPRI's Development Strategies and Governance Unit, Washington, DC, USA.

Jan Duchoslav is a senior research fellow with IFPRI's Development Strategies and Governance Unit, Lilongwe, Malawi.

Olivier Ecker is a senior research fellow with IFPRI's Foresight and Policy Modeling Unit, Washington, DC, USA.

Derek Headey is a senior research fellow with IFPRI's Development Strategies and Governance Unit, Colombo, Sri Lanka.

Charlotte Hebebrand is director of IFPRI's Communications and Public Affairs Unit, Washington, DC, USA.

Manuel A. Hernández is a senior research fellow with IFPRI's Markets, Trade, and Institutions Unit, Washington, DC, USA.

Kalle Hirvonen is a senior research fellow with IFPRI's Development Strategies and Governance Unit, Helsinki, Finland.

Naureen Karachiwalla is a research fellow with IFPRI's Poverty, Gender, and Inclusion Unit, Washington, DC, USA.

Nadim Khouri is an independent researcher on food security and climate-smart agriculture, Washington, DC, USA.

Oliver Kirui is a research fellow with IFPRI's Development Strategies and Governance Unit, Khartoum, Sudan.

Sikandra Kurdi is a research fellow with IFPRI's Development Strategies and Governance Unit, Dubai, United Arab Emirates.

Abdullah Mamun is a senior research analyst with IFPRI's Markets, Trade, and Institutions Unit, Washington, DC, USA.

Will Martin is a senior research fellow with IFPRI's Markets, Trade, and Institutions Unit, Washington, DC, USA.

Pavlo Martyshev is a researcher with the Center for Food and Land Use Research, Kyiv School of Economics (KSE Agrocenter), Kyiv, Ukraine.

Juneweenex Mbuthia is a research officer with IFPRI's Development Strategies and Governance Unit, Nairobi, Kenya.

Olga Nivievskyi is an assistant professor and vice president for Economics Education, with the Kyiv School for Economics, Kyiv, Ukraine.

Elsa Olivetti is an intern with IFPRI's Markets, Trade, and Institutions Unit, Washington, DC, USA.

Lensa Omune is a research officer with IFPRI's Development Strategies and Governance Unit, Nairobi, Kenya.

Edwin Ombui Oseko is an officer with the Fertilizer Promotion Unit, Kenya Ministry of Agriculture, Livestock, Fisheries, and Cooperatives, Abuja, Nigeria.

Valeria Piñeiro is acting head of the Latin American region and a senior research coordinator with IFPRI's Markets, Trade, and Institutions Unit, Washington, DC, USA.

Angga Pradesha is a senior research analyst with IFPRI's Development Strategies and Governance Unit, Washington, DC, USA.

Brendan Rice is a research analyst with IFPRI's Markets, Trade, and Institutions Unit, Washington, DC, USA.

Claudia Ringler is director of IFPRI's Natural Resources and Resilience Unit, Montreal, Canada.

Marie Ruel is a senior research fellow with IFPRI's Nutrition, Diets, and Health Unit, Washington, DC.

Yousra Salama is a former senior research assistant with IFPRI's Development Strategies and Governance Unit.

Carin Smaller is co-founder and executive director of the Shamba Centre for Food & Climate, Geneva, Switzerland.

Johan Swinnen is managing director of Systems Transformation, CGIAR, and director general of IFPRI, Washington, DC, USA.

Agustín Tejeda is Chief Economist of the Buenos Aires Grain Exchange, Buenos Aires, Argentina

James Thurlow is a senior research fellow with IFPRI's Development Strategies and Governance Unit, Washington, DC, USA.

Fousseini Traoré is a senior research fellow with IFPRI's Markets, Trade, and Institutions Unit, Dakar, Senegal.

Rob Vos is director of IFPRI's Markets, Trade, and Institutions Unit, Washington, DC, USA.

Made in the USA
Monee, IL
12 September 2026

62981015R00119